ADVANCES IN MAIZE SCIENCE

Botany, Production, and Crop Improvement

ADVANCES IN MAIZE SCIENCE

Botany, Production, and Crop Improvement

Ratikanta Maiti
Humberto González Rodríguez
Ch. Aruna Kumari
Sameena Begum
Dasari Rajkumar

First edition published 2022

Apple Academic Press Inc.
1265 Goldenrod Circle, NE,
Palm Bay, FL 32905 USA
4164 Lakeshore Road, Burlington,
ON, L7L 1A4 Canada

CRC Press
6000 Broken Sound Parkway NW,
Suite 300, Boca Raton, FL 33487-2742 USA
2 Park Square, Milton Park,
Abingdon, Oxon, OX14 4RN UK

© 2022 Apple Academic Press, Inc.

Apple Academic Press exclusively co-publishes with CRC Press, an imprint of Taylor & Francis Group, LLC

Library and Archives Canada Cataloguing in Publication

Title: Advances in maize science : botany, production, and crop improvement / Ratikanta Maiti, Humberto González Rodríguez, Ch. Aruna Kumari, Sameena Begum, Dasari Rajkumar.
Names: Maiti, R. K., 1938- author. | González Rodriguez, Humberto, 1959- author. | Aruna Kumari, C. H., 1972- author. | Begum, Sameena, author. | Rajkumar, Dasari, author.
Description: First edition. | Includes bibliographical references and index.
Identifiers: Canadiana (print) 20210136723 | Canadiana (ebook) 20210136804 | ISBN 9781771889520 (hardcover) | ISBN 9781774638255 (softcover) | ISBN 9781003104995 (PDF)
Subjects: LCSH: Corn.
Classification: LCC SB191.M2 M35 2021 | DDC 633.1/5—dc23

Library of Congress Cataloging-in-Publication Data

Names: Maiti, R. K. 1938- author. | González Rodriguez, Humberto, 1959- author. | Aruna Kumari, C. H., 1972- author. | Begum, Sameena, author. | Rajkumar, Dasari, author.
Title: Advances in maize science : botany, production, and crop improvement / Ratikanta Maiti, Humberto González Rodríguez, Ch. Aruna Kumari, Sameena Begum, Dasari Rajkumar.
Description: 1st edition. | Palm Bay, FL, USA : Apple Academic Press, 2021. | Includes bibliographical references and index. | Summary: "Advances in Maize Science: Botany, Production, and Crop Improvement offers a multi-pronged perspective on maize science, bringing together important recent research advances from several disciplines. The volume covers maize from origin to biotechnology. It provides an overview of recent world maize production along with technological advancements and green strategies in maize science. The authors cover the background of maize, its origin and domestication, ideotypes, botany, taxonomy, physiology of crop growth, methods of cultivation, production, nutritional functions, biotic and abiotic stress impacts, postharvest management and technology, maize grain quality, and advances in breeding and biotechnology, filling a gap in the literature of maize. The authors consider that productivity of maize is affected by several biotic and abiotic stresses that require a concerted interdisciplinary research. This book aims to be a comprehensive guide for maize scientists, agronomists, breeders, researchers, academicians, and students. It will guide researchers working in maize science and maize scientists to understand the relation between the disciplines and implementation of new methods and technology for crop improvement"-- Provided by publisher.
Identifiers: LCCN 2021006676 (print) | LCCN 2021006677 (ebook) | ISBN 9781771889520 (hardcover) | ISBN 9781774638255 (paperback) | ISBN 9781003104995 (ebook)
Subjects: LCSH: Corn--Research. | Corn--Growth.
Classification: LCC S191.M2 M358 2021 (print) | LCC S191.M2 (ebook) | DDC 633.1/5072--dc23
LC record available at https://lccn.loc.gov/2021006676
LC ebook record available at https://lccn.loc.gov/2021006677

ISBN: 978-1-77188-952-0 (hbk)
ISBN: 978-1-77463-825-5 (pbk)
ISBN: 978-1-00310-499-5 (ebk)

About the Authors

Ratikanta Maiti, PhD, DSc, was a world-renowned botanist and crop physiologist. He worked on jute and allied fibers at the former Jute Agricultural Research Institute (ICAR), India, and as a plant physiologist on sorghum and pearl millet at ICRISAT (International Crops Research Institute for the Semi-Arid Tropics). He also worked for more than 25 years as a professor and research scientist at three different universities in Mexico. He was also a Research Advisor at Vibha Seeds, Hyderabad, India, and as a Visiting Research Scientist at the Universidad Autónoma de Nuevo León, Facultad de Ciencias Forestales (School of Forest Sciences), Nuevo León, México. As the author of more than 40 books and about 500 research papers, he won several international awards, including an Ethno-Botanist Award (USA) sponsored by Friends University, Wichita, Kansas, the United Nations Development Programme; a senior research scientist award offered by Consejo Nacional de Ciencia y Tecnología (CONACYT), México; and gold medal from India 2008 offered by ABI. He was Chairman of the Ratikanta Maiti Foundation and chief editor of three international journals. Dr. Maiti died in 2019.

Humberto González Rodríguez, PhD, is a faculty member at the Universidad Autónoma de Nuevo León, Facultad de Ciencias Forestales (School of Forest Sciences), Nuevo León, México. He is currently working on water relations and plant nutrition in native woody trees and shrubs in northeastern México. In addition, his research includes nutrient deposition via throughfall, stemflow, and litterfall in different forest ecosystems. Dr. González teaches chemistry, plant physiology, and statistics. He has successfully guided over 68 theses and has handled over 10 research projects. Moreover, he has published many articles, books, and book chapters. He received his PhD in Plant Physiology from Texas A&M University under the guidance of Dr. Wayne R. Jordan and Dr. Malcolm C. Drew.

Ch. Aruna Kumari, PhD, is an Assistant Professor in the Department of Crop Physiology at Agricultural College, Jagtial, Professor Jayashankar Telangana State Agricultural University (PJTSAU), India. She has teaching experience at PJTSAU and research experience at varied ICAR institutes and at Vibha Seeds. She was the recipient of a CSIR Fellowship during her doctoral studies and was awarded a Young Scientist Award for Best Thesis Presentation on "In the National Seminar on Plant Physiology." She teaches courses on plant physiology and environmental science for BSc (Ag.) students. She has taught seed physiology and growth and yield and modeling courses to MSc (Ag.) students. She acted as a minor advisor to several MSc (Ag) students and guided them in their research work. She is the author of book chapters in four books. She is also one of editors of the book *Glossary in Plant Physiology* and an editor of six international books, including *Advances in Bio-Resource and Stress Management; Applied Biology of Woody Plants; An Evocative Memoire: Living with Mexican Culture, Spirituality and Religion;* Gospel of Forests; Advances in Cotton Science; Experimental Ecophysiology and Biochemistry of Trees and Shrubs; and Advances in Rice Science. She has published over 50 research articles in national and international journals.

Sameena Begum, a young researcher, has completed a BSc Agriculture with distinction in the year 2016 and MSc in Genetics and Plant Breeding with distinction in the year 2018 from the College of Agriculture at the Professor Jayashankar Telangana State Agricultural University, Hyderabad, India. During her master's degree program, she conducted research on combining ability, gall midge resistance, yield, and quality traits in hybrid rice (*Oryza sativa* L.) and identified two highly resistant hybrids. She has been an author of book chapters in two books, namely *Advances in Cotton Science* and *Experimental Ecophysiology and Biochemistry of Trees and Shrubs.* She was also an editor of a book on *Advances in Cotton Science* (Apple Academic Press). She has been an author of book chapters in three books: Experimental Ecophysiology and Biochemistry of Trees and Shrubs, Advances in Cotton Science and Advances in Genetics and Plant Breeding. She was one of the editors of the books: Advances in Cotton Science and Advances in Rice Science. She published five research publications in national and international journals from her research work.

Dasari Rajkumar, is a rice breeder. His career included work as a research associate and a botanist at Vibha Seeds Hyderabad, India, where he contributed with research on physiological aspects of drought and salinity tolerance in paddy, cotton, and vegetables. He also was a senior research associate at Bio Seed Research India, Hyderabad, where he specialized in rice, focusing on rice breeding and QTL mapping for drought, salinity tolerance, and trait introgression for biotic stress tolerance. Currently, he is working as a rice breeder in Neo Seeds India Private Limited, Hyderabad, India. His main area of research is on developing high-yielding drought and salinity tolerant rice hybrids and varieties of various maturing segments and grain types across India. Apart from this, he is working on biotic stress tolerance in paddy at various hotspots across India and a trait introgression program for biotic stress tolerance. He has co-authored three books and published 15 research papers in national and international journals. He completed his master's degree in Botany with a specialization in Applied Plant Physiology and Molecular Biology from Osmania University, India.

Contents

Abbreviations

AAG	accelerated aging germination
ABA	abscisic acid
AEZ	agro-ecological zones
ALAB	amylolytic lactic acid bacteria
APSIM	Agricultural Production Systems Simulator
APX	ascorbate peroxidase
ASI	anthesis-silking interval
BR	brace roots
CCCM	Canadian Climate Center Model
CIM	composite interval mapping
CR	crown roots
CS	crop susceptibility
CT	conventional tillage
CTG	cold test germination
CSM	cropping system model
DAP	days after pollination
DB	dry basis
DEA	Data Envelopment Analysis
DH	doubled haploid
DL	dry land
DTI	drought-tolerance index
EDD	extreme degree days
EM	excessive moisture
EN	ear number
ETEC	enterotoxigenic *Escherichia coli*
FSZT	flat sowing with zero tillage
FYM	farmyard manure
GBLUP	genomic best linear unbiased prediction
GBS	genotyping by sequencing
GCA	general combining ability
GCMs	general circulation models
GDH	glutamate dehydrogenase
GFDL	Geophysical Fluid Dynamics Laboratory

GFR	grain-filling rate
GHG	greenhouse gases
GIS	geographic information system
GLS	gray leaf spot
GM	genetically modified
GM	grain moisture
GPX	guaiacol peroxidase
GR	glutathione reductase
GS	glutamine synthetase
GY	grain yield
HIR	haploid induction rate
HS	head smut
HSPs	heat shock polypeptides
HSW	hundred seed weight
HU	heat unit
GWAS	genome-wide association studies
IAPM	integrated agronomic practices management
IHT	ideal harvest time
IRT	infrared thermometer
IS	intermediate stress
IWUE	irrigation water use efficiency
KN	kernel number
LAI	leaf area index
LD	linkage disequilibrium
LOX	lipoxygenase
LR	lateral roots
MAF	minor allelic frequency
MAS	marker-assisted selection
MBP	myelin basic protein
MLM	modified location model
MMV	maize mosaic virus
MODIS	moderate resolution imaging spectroradiometer
MRD	Modified Roger's distance
MSD	maize streak disease
MSV	maize streak virus
NAM	nested association mapping
NCLB	northern corn leaf blight
NCS	normalized crop susceptibility

NAT	nitrogen rate testing
NILs	near-isogenic lines
NIR	near-infrared
NLB	northern leaf blight
NR	nitrate reductase
NUE	nitrogen use efficiency
OLS	Ordinary Least Square
PA	phosphatidic acid
PAR	photosynthetically active radiation
PCA	principal component analysis
PLD	phospholipase D
PLFA	phospholipid fatty acid
PM	physiological maturity
PNUE	photosynthetic N-use efficiency
QPR	quality protein maize
QTL	quantitative trait loci
RFR	red:far-red ratio
RFLP	restriction fragment length polymorphism
RGA	resistance gene analog
RKHS	Reproducing kernel Hilbert spaces
RL	root length
RUM1	rootless with undetectable meristems 1
RZWQM	Root Zone Water Quality Model
SA	surface area
SCA	specific combining ability
SCR	Southern corn rust
SDI	stress-day index
SG	standard germination
SMB-C	soil microbial biomass carbon
SMB-N	soil microbial biomass nitrogen
SNP	single-nucleotide polymorphism
SOD	superoxide dismutase
SS	severe stress
SSRs	simple sequence repeats
VPD	vapor pressure deficit
WUE	water-use efficiencies
WW	well-watered
XET	xyloglucanendo-transglycosylase

Preface

Maize (*Zea mays* L.) is also called corn. It is the world's number 1 grain in terms of production and number 3 as a dietary staple after rice and wheat, accounting for 5% of the world's human caloric intake.

Maize was first domesticated by native peoples in Mexico about 10,000 years ago. It is mainly used as a food source, important raw material in industry, biofuel, for forage, and also for animal feed. It is the most important cereal grain that can be grown worldwide in a wide range of environments because of its greater adaptability.

Maize plays an important role in most of a country's economy, especially in developed and developing countries. Based on the importance of the crop, farmers, students, researchers, and scientists need to understand the role, nature of the crop, and advancements in research for best cultivation methods, effective utilization of resources, and operations for attaining higher yields, thus achieving higher productivity.

During recent decades, tremendous progress and innovations have been made in all fields of maize science, including super sweet and sweet corn, baby corn, and double haploid maize, and multiple disease resistance including molecular biology—all these for attempting to increase maize's desirable traits and to increase production of maize to meet the world's demands and hunger. From the above point of view, the authors decided to provide a resource of complete information and research literature of several disciplines on maize in the form of this book as a guide for students, teachers, researchers, as well as scientists.

The authors provide information on all aspects of several disciplines of maize along with recent literature, together under one umbrella, namely *Advances in Maize Science: Botany, Production, and Crop Improvement*. This book attempts to bring together recent advances in different disciplines of maize science. This book covers most of the aspects of maize starting from background, production, origin to domestication, ideotype, botany, physiology of crop growth and productivity, abiotic and biotic factors affecting crop productivity, methods of cultivation, postharvest management, grain quality analysis, to food processing, improvement of maize crop, research advancements in breeding and biotechnology up to 2019. We need to be concerned that the productivity of maize is affected by several biotic and abiotic stresses, which requires concerted interdisciplinary research. The content of each and every chapter is described extensively and enriched with recent research literature.

This mode of presentation will help the students of undergraduate, graduate, academicians, and teaching faculty to gain the knowledge and understand the crop. Especially, this book guides researchers working on maize and maize scientists to understand the relation between several disciplines and implementation of new methods and technology covered in recent literature for maize crop improvement for higher productivity. A multi-pronged approach needs to be directed to increase maize productivity to meet the world's demands and hunger.

The authors strongly believe that libraries of schools or colleges of undergraduate, graduate, postgraduate, research institutes of public and private sectors must have this book. It also will occupy a distinctive place in libraries for its versatile content, extensive descriptions, and enriched research literature of recent advancements. This book is effectively helpful in the aspect of gaining knowledge, gathering subject matter, quick reference for teaching staff, professors, research community, and maize scientists.

Some problems were faced during the course of writing this book, but the authors' dedication and determination played a major driving force in overcoming the problems and successfully completing the book.

The authors sincerely thank Apple Academic Press for publishing this book.

ACKNOWLEDGMENT

Authors acknowledge the following public and private institutes for playing important role in society and providing an excellent platform for their carriers and continuous support in the area of research and publications:

1. ICRISAT, Patancheru, Telangana State, India
2. Jute Agricultural Research Institute, Barracpore (I.C.A.R.), Kolkata, West Bengal, India
3. Universidad Autónoma de Nuevo León, Facultad de Ciencias Forestales (School of Forest Sciences), Nuevo León, México

4. Department of Botany, Osmania University, Hyderabad, Telangana State, India
5. Research and Development Centre, Neo Seeds India Private Limited, Hyderabad, Telangana State, India
6. Vibha Agro Tech Pvt. Ltd, Madhapur, Telangana State, India

The authors thank Dasari Rajkumar for his courtesy in supplying and organizing original photographs of the maize plant.

CHAPTER 1

Background and Importance of Maize

ABSTRACT

Maize is becoming an ideal staple food owing to its low costs of production and high consumption rates. Additionally, it can be processed into several products which generate an additional source of income for maize farmers, maize processors, and distributors. This chapter presents the importance of maize at the global level and discusses a brief outline of research advances in various aspects of the maize crop.

1.1 IMPORTANCE OF MAIZE

Maize (*Zea mays*) is commonly referred to as corn. Research evidences indicate that it had its origin in Central Mexico, 7000 years in the past from a wild grass. It has been converted into a good source of food by Native Americans. It is widely grown throughout the world. Some of the major constituents of maize grain are starch (72%), protein (10%), and fat (4%). This cereal has a capacity of supplying energy of approximately 365 Kcal/100 g. The top most maize producing countries are the United States, China, and Brazil, which produce around 563 of the 717 million metric tons/year. Maize can easily be processed into a multiplicity of food and industrial products. Some of the commonly used products are starch, sweeteners, oil, beverages, glue, industrial alcohol, and fuel ethanol. In the preceding 10 years, its use for the production of fuel has considerably increased. This fuel production in the United States accounts to 40% of the maize production where most of the maize grain produced is used in the production of ethanol. Thus, the ethanol industry has a larger share of maize requirement. It has also its utility for animal and poultry feed because of its demand and competition, higher prices for maize may increase this demand competition. Maize is becoming an ideal staple food owing to its low costs of production and high consumption rates particularly by people suffering from micronutrient deficiencies and health problems (Ranum et al., 2014).

Types of maize/corn

Flint corn: This is a variant of maize, which has a hard outer layer and low water content. It is multicolored and is utilized for decorations in the United States.

Dent corn: Dent corn has high soft starch content and is commercially cultivated for grain and fodder.

Pod corn: Also known as wild maize is the most primitive variety of maize and recognized best as the progenitor of corn.

Popcorn: A variant of corn that swells and puffs on heating. When heated, the kernel ruptures and permits the kernel content to expand, cool, and finally set into popcorn.

Flour corn: Flour corn has a soft endosperm and utilized mainly to prepare corn flour.

Sweet corn: Sweet corn has a high sugar content and is generally considered a vegetable. Sweet corn is best when consumed fresh or canned to preserve its freshness (Farm rowdy).

Economic Importance

Maize is a rich source of vitamins, minerals, and dietary fiber. As many small-scale farmers are engaged in maize farming, it makes it a cheap source of vitamins and minerals for rural people.

Maize can be processed into several products which generate an additional source of income for maize farmers, maize processors, and distributors. Some of the processed products that can be made from maize comprise:

- Corn Starch: Is utilized as a thickener for liquid food, is the key ingredient in biodegradable plastic, a component that can be utilized to substitute talc in body powder, and is also applied by dry cleaners to maintain clothes firm.
- Oil: Oil obtained by squeezing the corn germ is mostly used to prepare crunchy, sweet popcorns. Further, it can be utilized to produce margarine and in the making of soap, cosmetics, etc.

- Glue: Corn germ can be processed to make industrial glue stronger. This reduces the cost of industrial glue production.
- Ethanol: By distilling corn, an alcohol called ethanol can be produced which can be combined with gasoline and utilized in powering vehicles. Gasoline generally contains ethanol in the ratio 10:90 (10—ethanol, 90—gasoline) to oxygenate the fuel and decrease air pollution. Ethanol is a valuable solvent that can be utilized in domestic products like paints and varnish.
- Ethanol can be used to kill microorganisms and a common ingredient in cosmetics, beauty products, and hand sanitizers. Its capacity to effectually kill microorganisms makes it an excellent preservative.

Maize can be cooked, roasted, or blended and utilized in delicacies like fried rice, etc. Blended corn can be utilized for pancakes, baby food, and baking.

Maize can be processed into diverse products and more employment opportunities will be generated. This in turn causes a fall in the cost of purchasing maize and maize products. Because of all these more people will be able to afford good, quality food.

1.2 ETYMOLOGY

The determination of time and path of the maize introduction into West Africa could cast light on primary records about the area. This would assist in clarifying several questions of the indigenous cultures interactions with each other along with intrusive cultures, and it can propose a component of absolute dating into the chronology of archaeological deposits in the area. But, most of the methods for dating archaeological material in absolute terms cannot be used in the study of the ethno history of West Africa. The maize cob was extensively utilized in West Africa as a decorative roulette on pottery, generally on the coarser wares. So, the presence or absence of such decoration was used as a horizon in digging archaeological sites in West Africa. Furthermore, botanists have examined the genetics of maize in West Africa in detail, so that a statistical study of maize imitations on pottery, could allow the identification and relative location in time of variations in the maize population grown at a given site, and thus help to date the pottery (Willett, 1962).

Mexico is the center of domestication and diversity of maize and its ecology has been investigated in Mexico for several decades. Although the wide summaries of diversity and dynamics of native maize populations were identified at the farm and national levels, these topics were not well known at the landscape level. The recent research suggests that apart from environmental factors which are the primary forces influencing the diversity of the species in Mexico, current social origin, such as community of residence or ethnolinguistic group, also affects the maize population structure at more local levels. A landscape viewpoint can help to find out whether these social factors work in a constant manner through diverse environments. Brush and Perales (2007) used the data from the Chiapas Case study to exemplify the role of ethnicity in understanding the maize diversity ecology in Mexico. They found that the environmental variations are significant in ascertaining the general pattern of maize diversity across the Chiapan landscape but social origin has a substantial influence on maize populations in all environments.

Though the correlation between ethnolinguistic diversity and crop diversity has been recorded, much organized study has not concentrated on the role of culture in shaping crop diversity. In this study, Perales et al. (2005) assessed the dispersal of maize (*Zea mays*) types among communities of two groups, the Tzeltal and Tzotzil. The results indicated that maize populations were separated as stated by ethnolinguistic group. But ethnolinguistic origin-based separation has not been clearly shown by an analysis of isozymes. A reciprocal garden experiment showed that the maize is well adapted to its environment; however, the Tzeltal maize occasionally outyields Tzotzil maize in Tzotzil environments. Due to the closeness of the two groups and selection for yield, it is assumed that the superior maize would dominate both groups' maize populations, but it was found that such domination is not the case. Thus, in relation to landrace differentiation, they discussed the role of ethnolinguistic diversity in determining social networks and information exchange.

1.3   MAIZE-GROWING ENVIRONMENTS

The conditions in which maize (*Zea mays* L.) germplasm originated and in which it is assessed can significantly influence outcomes from germplasm assessments, hence affecting where the germplasm will ultimately be utilized. The normally used adaptation classifications such as temperate, tropical, subtropical, and highland are inaccurate. So, Pollak and Corbett (1993) applied multivariate statistical techniques to spatial geographic information system (GIS) datasets of agroclimatic data group from related maize-growing regions in Mexico and Central America. They used these groups to improve the mega-environments developed by CIMMYT maize breeders to help manage their germplasm. On the basis of each year's monthly data, mean monthly temperatures and precipitation, total precipitation, and mode of the elevations in the grid, the variables such as mean maximum and minimum monthly air temperatures, absolute maximum and minimum air temperatures were analyzed. The cluster analysis was applied on 7 months of growing season data (April through October) to get 25 groups. Then, these 25 groups were categorized into 10 maize ecologies analogous to CIMMYT's mega-environments. The ecologies comprised three lowlands, three highlands, two subtropical, and two transitional from subtropical to highland. This technique would significantly help in categorizing and utilizing northern Latin America's large amount of diverse maize germplasm.

Spedding et al. (2004) studied the long-term effect of tillage and residue management on soil microorganisms in a sandy loam to loamy sand soil of southwestern Quebec, by growing maize monoculture. No till, reduced tillage, and conventional tillage with crop residues either taken out from (−R) or retained on (+R) experimental plots were applied as treatments. At two depths viz 0–10 and 10–20 cm, soil microbial biomass carbon (SMB-C), soil microbial biomass nitrogen (SMB-N) and phospholipid fatty acid (PLFA) contents were assessed four times during the growing season. The magnitude of time influence was greater than those accredited to tillage or residue treatments. The SMB-N exhibited a high response to the post-emergence application of mineral nitrogen, whereas SMB-C exhibited insignificant seasonal change (160 µg C g^{-1} soil) and PLFA analysis has shown a rise in fungi and total PLFA all over the season. PLFA profiles revealed a better difference between sampling time and depth than among treatments. In comparison, the +R plots exhibited a more pronounced residue effect with increased SMB-C (61%) and SMB-N (96%). These findings demonstrated that while evaluating soil quality on the basis of soil microbial components the seasonal variations in soil physical and chemical conditions should be taken into account.

Hirasawa and Hsiao (1999) conducted a study to measure the maize leaf photosynthetic rate over diurnal courses on cloudless days with the leaf held perpendicular to the sunlight.

For the study maize was cultivated in the high-radiation arid summer environment of Davis, California. The maximum leaf photosynthesis was recorded in the late morning on days of high atmospheric vapor pressure deficit (VPD) and then reduced slowly as the day advanced, although the soil was well irrigated. In the measurement chamber when CO_2 concentration was increased to about 1000 μmol mol^{-1}, the higher photosynthesis was recorded in the afternoon than in the morning. However, a significant difference was not observed in the curves of photosynthetic rate (A) vs intercellular CO_2 concentration (C_i) for the morning and afternoon. Therefore, photosynthetic capacity was alike for the two periods and there was no proof of photo inhibition by the high photosynthetic photon flux density at noon. Additionally, C_i and photosynthetic rates A quantified over different photon flux densities were lower in the afternoon than in the morning. These results specify that epidermal conductance (mostly stomatal) limits the photosynthetic rates A at noon and early afternoon. On a day with low VPD, midday depression in these results specifies that epidermal conductance (mostly stomatal) limits the photosynthetic rates A and which were not marked for the well-irrigated plants. However, in plants without irrigation and at lower midday water potential the reduction in conductance and photosynthetic rates A was much clearer, initiating late in the morning. Thus, they concluded that the low leaf water potential affected by high transpiration rates causes midday reduction in conductance and photosynthetic rates A.

To decrease agricultural water use in water deficit areas, a study on the response of crop to deficit irrigation is essential. So, to determine the response of maize (*Zea mays* L.) to deficit irrigation Farré and Faci (2009) conducted two field trials on a loam soil in northeast Spain. All possible combinations of full irrigation or limited irrigation were applied as treatment in the three phases, that is, vegetative, flowering, and grain filling. Then, the interval between irrigations was increased to apply limited irrigation. Water status of soil, crop growth, above-ground biomass, yield, and its component traits were evaluated. Findings revealed that flowering is the most sensitive stage to water scarcity and water deficit at this stage results in higher biomass, yield, and harvest index reductions. Around flowering, treatments with deficit irrigation had significantly lower average grain yield than the well-irrigated treatments and the irrigation water use efficiency (IWUE) was greater in fully irrigated treatments. During the grain filling phase, the deficit irrigation or higher interval between irrigations had no significant effect on crop growth and yield. These results indicate that in maize relatively high yields can be maintained if slight water deficits bring about by increasing the interval between irrigations were restricted to phases other than the flowering stage.

In maize water deficit at tasseling and silking causes significant reductions in grain number. In order to elucidate the causes of reductions in kernel number under the mild stresses typical of humid regions, more information on the responses of crop to water supply is needed. A field experiment was undertaken by Otegui et al. (1995) to measure crop evapotranspiration, E_c, and its association with shoot biomass production, grain yield, and kernel number. To create differences in evaporative demand the experiment was conducted with two sowing dates (6 weeks apart). Then to create a 40-day period of reduced water supply during silking plastic covers were laid on the ground of water-deficit plots. Water deficit affected plant height, maximum leaf area index, and shoot biomass. Shoot biomass accumulation was associated with E_c, but the water-stress treatments had greater water-use efficiencies (WUE). During the treatment period, grain yield was found to be associated to kernels m^{-2} ($r = 0.88$; 6 d.f.), and both grain yield and kernels m^{-2} were correlated with E_c. Further, when fresh pollen was applied to late appearing silks, the number of kernels per ear did not increase indicating that ovaries which failed to expose their silks synchronously with pollen shedding were deleteriously influenced by water deficit.

On the basis of GIS from the views of climatology, geography, disaster science, and environmental science Zhang (2004) described a method for risk analysis and estimation of drought disaster to agricultural production in the maize-growing area of Songliao Plain of China. In order to study and measure associations among the variation of maize yield and agrometeorological disasters, and to assess the effects of drought disaster crop yield–climate analysis and regression analysis were used. The data on historical climate, crop yield, crop sown area, crop damaged area, and crop loss was collected from 41 maize-producing districts of Songliao Plain. It was found that among all agrometeorological disasters, drought occurs with the greatest frequency, covers the largest area, and brings about the greatest loss to agricultural production, and the economy in the region. The negative value years of the maize yield variation because of agrometeorological disasters accounted for 55%, of which 60% was affected by drought. The negative values of maize yield variation were significantly and positively associated with drought affected areas, indicating the adverse effects of drought on maize production are comparable to the magnitude of damage caused by drought disasters. On the basis of drought disaster risk extent to maize, the Songliao Plain was separated into four subregions: high risk zone, medium risk zone, low risk area, and slight risk zone by applying fuzzy cluster analysis. It revealed that the risk degree of drought disaster in the Songliao Plain rises progressively from south to north and from east to west. The results of this study can be used as a base to develop strategies to lessen drought and reduce losses and ensure sustainable agricultural development.

Cross and Zuber (1972) undertook an evaluation of different thermal unit formulas to find better ways of assessing relative maturity variances in corn (*Zea mays* L.). Around 22 different methods of thermal units were computed for their capacity to elucidate variation in flowering dates by using data from six plantings of corn over a 2-year period. Hourly as well as daily temperature data were utilized in the equations. On the whole, the daily measurements seemed to be roughly as accurate as the hourly measurements. The best thermal units equation for predicting flowering dates used a base temperature of 10°C (50°F) and an optimal of 30°C (86°F). Therefore, to calculate high temperature stress, the excess temperature above 30°C should be deducted.

The emission of greenhouse gases (GHG) from unintended land-use change activated by crop-based biofuels has been at the center of attention in the dispute over the role of biofuels in climate strategy and energy security. Hertel et al. (2010) analyzed the emission of GHG during the production of ethanol from maize in the United States. The factorization of market-mediated responses and by-product utilized in the analysis reduced cropland change by 72% from the land utilized for the ethanol feedstock. Subsequently, the related GHG emission assessed in the background was 800 g of carbon dioxide per megajoule (MJ); in 30 years of ethanol production, or approximately one fourth of the other available assessments of emissions were attributed to variations in unintended land use. However, 800 g is sufficient to negate the advantages that corn ethanol has on global warming, thus reducing its probable contribution in the context of California's Low Carbon Fuel Standard.

To increase the understanding of the nitrogen use efficiency genetic basis in maize (*Zea mays*), Hirel et al. (2001) developed a quantitative genetic method by linking metabolic functions and agronomic traits to DNA markers.

The maize recombinant inbred lines with known agronomic performance were evaluated for physiological traits like nitrate content, nitrate reductase (NR), and glutamine synthetase (GS) activities. A significant genotypic variation was observed among these traits and a positive association was noticed among nitrate content, GS activity and yield, and its components. Conversely, NR activity had a negative association with other traits. These findings reveal that the capability of maize to store nitrate in their leaves during vegetative growth and to effectively remobilize this stored nitrogen during grain filling is responsible for improved productivity in maize genotypes. They searched and located the quantitative trait loci (QTL) for different agronomic and physiological traits on the genetic map of maize. Concurrences of QTL for yield and its component traits with genes encoding cytosolic GS and the subsequent enzyme activity were identified. It was found that the GS locus on chromosome 5 is a useful candidate gene that can partly elucidate differences in yield or kernel weight. Since at GS locus concurrence of QTLs for grain yield, GS, NR activity, and nitrate content were also detected, they assumed that accumulation of leaf nitrate and the reactions catalyzed by NR and GS are co-regulated and are the main elements regulating nitrogen use efficiency in maize.

A study has been undertaken by Eakin (2000) to investigate the approaches employed by smallholder farmers in Mexico to deal with the effects of climatic variability, and in what way seasonal climate forecasts may help these farmers in alleviating climatic risk. Knowing that the small-holder farmer's takes decisions based on their political-economic conditions in which they work, the article deliberates in what way agricultural policy in Mexico influences the weakness of small-scale producers and prevents their ability to utilize climatic forecasts for their benefit. Initially, the literature on smallholder adaptation in Mexico was reviewed, and then policy and institutional problems influencing adaptation at the farm-level were discussed. By taking the situation of small-scale maize producers in Tlaxcala, Mexico, as an example, the article then debates that political-economic uncertainty counterweights climatic variability as a determining factor of the production approaches of small-scale producers. In these conditions, the farmers are improbable to use novel seasonal climate forecasts.

The common problems that come across during the growing season are water deficits, suboptimum temperatures, and low levels of solar radiation. The impact of these varying weather conditions on crop production can be evaluated using crop simulation models. Tojo Soler et al. (2007) evaluated the cropping system model (CSM)-CERES-Maize for its capacity to simulate growth, development, grain yield for four diverse maturity maize hybrids cultivated off-season in a subtropical region of Brazil. Further, they studied the effect of various planting dates on the performance of maize under rainfed and irrigated conditions. The CSM-CERES-Maize assessment has shown that the model can simulate phenology and grain yield for the four hybrids precisely, with less than 15% of standardized RMSE. The planting date analysis revealed that in all hybrids a delayed planting date from February 1 to April 15 reduced the average yield by 55% and 21% for the rainfed and irrigated conditions, respectively. The yield forecasting analysis showed that a precise yield prediction could be given at around 45 days before harvesting for all four maize hybrids. These findings have potential for farmers and decision makers, as they could get precise yield forecasts before the final harvest. But, to make practical decisions for stock management of maize grains, this method should be developed for different locations. As the breeder's releases new cultivars, future model evaluations may be required.

1.4   ROLE OF MAIZE CROP IN HUMAN CIVILIZATION

Molecular evidence showed that the maize wild ancestor is currently native to the seasonally dry tropical forest of the Central Balsas watershed in southwestern Mexico. Ranere et al. (2009) reported archaeological explorations in a region of the Central Balsas in Guerrero state that exhibit an extended sequence of human occupation and plant use getting back to the early Holocene. Well-stratified buildups and a stone tool collection of bifacially flaked points, simple flake tools, and several hand stones and milling stone bases radiocarbon dated to 8700 calendrical years B.P were found in one of the excavated sites, Xihuatoxtla Shelter. As reported earlier, starch grain and phytolith deposits from the ground and chipped stone tools, along with phytoliths from directly related sediments, provided proofs for maize and domesticated squash around 8700 calendrical years B.P. The radiocarbon determinations, stratigraphic reliability of Xihuatoxtla's deposits, and the stone tool collection characteristics related with the maize and squash residues indicate that these crops were primary Holocene domesticates. In this region of Mexico, small groups of cultivators seem to have been involved in early agriculture who shift their settlements seasonally and engaged in various subsistence activities.

Staller and Thompson (2001) reported the existence of maize cob phytoliths in pottery residues from the late Valdivia Period site of La Emerenciana in their article. Pearsall (2002) made valuable additions to this proof for geographically widespread use of maize during the Valdivia period (Early Formative, 4500–2000 BC calibrated). The statements on the Late Valdivia La Emerenciana occurrence support the initial use of maize in Ecuador, while the use of maize in ceremonial instead of domestic during the Valdivia period were speculative. Thus, they concluded that the maize was known from the start of the Valdivia period, in domestic contexts, at the Real Alto site, and, the root crop cultivation assisted the fluorescence and extension of Valdivia culture.

When the ceramics originate from comparatively uniform archaeological contexts, the patterns of maize use can be effectively differentiated by analyzing absorbed organic residues in ceramics using the bulk stable carbon isotope method. This method is rapid and economical than the more frequently used compound-specific stable carbon isotope analysis. Furthermore, the bulk stable carbon isotope method can ascertain the presence of C4 plant carbon in samples with degraded organic compounds. Seinfeld et al. (2009) collected 24 samples of ceramic sherds from an Early Franco Period feasting deposit (ca. cal 650 BC) at the Olmec site of San Andrés, La Venta, Tabasco, Mexico. Then, using bulk stable carbon isotope analysis, the maize patterns were differentiated among these 24 samples. A comparison of the $\delta^{13}C$ results from different ceramics samples revealed that proportionally more maize was utilized in luxury beverage service wares than in utilitarian vessels, indicating that maize-based beverages were dominant in Mexico.

1.5 WORLDWIDE UTILIZATION

Researchers often assume maize as the staple food of Chavín civilization and some even debate that this crop offered the significant impetus for its development. A stable carbon isotope study allows an assessment of these assumptions. The maize is the only C_4 cultigen eaten in pre-Hispanic Peru, leaves a clear imprint on the bone chemistry of its consumers that permits the evaluation of its comparative significance in the diet. Examination of osteological samples from Chavín de Huántar and Huaricoto has shown that though maize was consumed, C_3 foods like potatoes and quinoa were major nutritional intakes during the development and highpoint of high-land Chavín civilization (ca. 850–200 B.C.). Burger and Van der Merwe (1990) explored the probable causes for the secondary significance of maize in the Early Period/Early Horizon subsistence systems of Chavín civilization.

Though archeologists have the ability to trace varying food usage in the archeological record, they have not used food systems in the social and political change study. To do so, consciousness of food meanings should be gained, which can later light up the planned usage of a specific food in the making of associations of need and prestige. Archeological proof from the central Andes of Peru shows that the role of maize transformed between A.D. 500 and 1500, shifting from a simply boiled cooking item, to a more composite symbolic food, converted via grinding and brewing into beer, with expanded political meanings. This transformation in maize processing and feeding arose at a time of intensified political and social pressures. Hastorf and Johannessen (1993) proposed that the change in maize usage reflected and contributed to new political dynamics, signifying how food habits can update archeologists about previous social and political systems.

The nixtamal is heat- and alkali-treated maize dough from which an acid beverage called Pozol is obtained on natural fermentation. During nixtamalization, the concentration of mono- and disaccharides from maize gets condensed and only starch remains available as the main carbohydrate for lactic acid fermentation. So as to, offer some base to know the role of amylolytic lactic acid bacteria (ALAB) in fermentation, Díaz-Ruiz et al. (2003) determined the diversity and physiological characteristics of ALAB. With phenotypic and molecular taxonomic methods, forty amylolytic strains were characterized. Then using ribotyping four different biotypes were differentiated of which *Streptococcus bovis* strains were found to be predominant. *S. bovis* strain 25124 exhibited very low amylase yield relative to biomass (139 U g [cell dry weight]$^{-1}$) and specific amylase production rate (130.7 U g [cell dry weight]$^{-1}$ h^{-1}). Further, *Streptococcus macedonicus*, *Lactococcus lactis*, and *Enterococcus sulfurous* strains were detected.

Loyola-Vargas and de Jimenez (1984) carried out a study to measure the aminative (NADH) and deaminative (NAD$^+$) glutamate dehydrogenase

(GDH) activities at different periods of time in maize calli, roots, and leaves homogenates. They exposed the calli and plantlets of maize (*Zea mays* L. var Tuxpeño 1) to specific nitrogen sources. The nitrogen sources exhibited different effects on the tissues tested. It was found that glutamate, ammonium, and urea inhibit GDH activity in callus tissue. Further, different activity ratios of amination and deamination reactions were observed under different nitrogen sources. Ammonium and glutamine enhanced the GDH-NADH activity in roots, while the same metabolites inhibited the GDH-NADH activity in leaves. Thus, the study findings suggest that each tissue has its specific GDH isoenzymes or conformers and their activities vary with the nutritional requisites of the tissue and the state of differentiation.

A flat cake (tortilla) prepared from alkali-treated maize is a traditional staple food of the Guatemalan people and a rich source of dietary calcium. Krause et al. (1992) conducted a cross-sectional study to examine rural-urban differences in tortilla size and limed maize usage in 60 houses of Kekchi indigenous people from three rural, two semi-urban, and one low-income urban community of Guatemala. When compared to semi-urban and urban tortillas, the average weight ± SEM of tortillas from rural areas was significantly greater. The size of tortillas made for sale was significantly less than those made for home consumption. This shows that the variances in tortillas may confuse quantitative assessment of tortilla consumption. It was found that semi-urban households prepare more limed maize per person than rural households, while the consumption of tortilla was less by urban than semi-urban adult women. A socioeconomic status indicator, housing quality had a significant and positive correlation with the amount of limed maize prepared per person. When the age index was utilized as a covariate for the household, a significant and inverse correlation was observed between the number of household members and the amount of limed maize prepared per person. This indicates that the nutrient requirements of members of large-sized households are yet to be met.

The northern leaf blight caused by *Setosphaeria turcica* is a major disease that affects maize. Dingerdissen et al. (1996) undertook a study to identify the QTL involved in the resistance of maize to this disease. Total 121 $F_{2:3}$ lines obtained from a cross between Mo17 (moderately resistant) and B52 (susceptible) were mapped to detect the QTLs. Based on marker data noted in an earlier study, a linkage map covering 112 RFLP loci with a mean interval length of 15 cM was made. Field tests were conducted by artificially inoculating the pathogen at three sites in tropical mid- to high-altitude regions of Kenya. Host plant response was evaluated in terms of incubation period, disease severity (five scoring dates), and the area underneath the disease progress curve (AUDPC). High heritability was recorded for all the traits. For the incubation period, QTLs were detected on chromosomes 2S and 8L. The QTLs associated with disease severity and AUDPC were located in the assumed centromeric region of chromosome 1 and on 2S, 3L, 5S, 6L, 7L, 8L, and 9S. The gene increasing latent period was located on 2S and found to be related with decreased disease severity of juvenile plants. All the QTL except QTL detected on chromosomes 3L, 5S, 7L, and 8L were affected by a large genotype x environment interaction. For both resistance and susceptibility, partially dominant gene action was predominantly recorded. Specific QTL elucidated 10 to 38% of the phenotypic variation of the traits. Except QTL on chromosomes 1, 6, and 9, all QTL were donated by the resistant parent Mo17. On chromosome 8L a QTL located to the similar region as the major race-specific gene *Ht2* favoring the allelic qualitative and quantitative resistance genes hypothesis.

In northern Guatemala, the typical Maya civilization was concentrated in lowlands and collapsed inexplicably in the ninth century AD. The left over lands were agriculturally rich carved without metal implements from a tropical rain forest, lands that had been cultivated with increasing intensity for 6–16 centuries. The Maya civilization apparently relocated in highlands to the south or less productive dry lowlands to the north.

In the tropics, major diseases and pests of maize were evaluated for their comparative importance in and nearby the Petén vs the highlands, and the viruses were emphasized. The plant hopper-borne virus, maize mosaic virus (MMV) was found to be the main reason for sustained crop failure of maize. Maize and teosinte are the only known hosts for the virus. Thus the disease has been severe only in maize grown as a monoculture in wet or irrigated tropics (e.g., Caribbean Islands, Venezuela, Hawaii, Tanzania, Australia). It was found that the gene Mv is the only known resistance source for this disease that provides a high level of resistance but not immunity. The Mv gene was found to be present in all seven of the races of maize grown in the Caribbean, but not in the primitive Mexican or Central American races.

It was reported that the MMV might have originated in northern South America at or about the time maize was brought into the Caribbean by the Arawak. In this region, the sympatric origin or selection in maize for the Mv resistance mutant was expected to cause its integration in all seven Caribbean maize races. Around the eighth century, it was estimated that viruliferous leafhoppers were blown from the Caribbean into the Petén leading to an epidemic of disease in susceptible maize races such as Nal-Tel and Tepecintle, cultivated by the Petén Maya. Continuous failure of maize production because of MMV would have distinguished areas of intensive maize cultivation, mainly where it was year-round. The disease would have been less serious in areas with a long dry season, such as north Yucatán and it would not have arisen in the highland areas to the south and west, areas to which surviving Maya probably migrated (Brewbaker, 1979).

Soares et al. (2012) described three new aflatoxin-producing species viz *Aspergillus mottae*, *A. sergii* and *A. transmontanensis*. These three species belongs to the *Aspergillus* section *Flavi* and were separated from Portuguese almonds and maize. Here, they studied morphology, extrolite production and DNA sequence data of these three isolates to describe new species. Phylogenetic analysis revealed that *A. transmontanensis* and *A. sergii* forms a clade with *A. parasiticus,* while *A. mottae* shares a most recent common ancestor with the *A. flavus* and *A. parasiticus* clade.

1.6 CONCLUSIONS

Maize is becoming a staple food and has great demand at the global level. It is a rich source of vitamins, minerals, dietary fiber, and can be processed into diverse products and more employment opportunities will be generated. Additional studies need to be directed to utilize maize and its products.

KEYWORDS

- **maize**
- **processed products**
- **industrial alcohol**

REFERENCES

Brewbaker, J. L. Diseases of Maize in the Wet Lowland Tropics and the Collapse of the Classic Maya Civilization. *Econ. Bot.* **1979,** *33,* 101–118.

Brush, S. B.; Perales, H. R. A Maize Landscape: Ethnicity and Agro-Biodiversity in Chiapas Mexico. *Agric. Ecosyst. Environ.* **2007,** *121,* 211–222. https://doi.org/10.1016/j.agee.

Burger, R. L.; Van der Merwe, N. J. Maize and the Origin of Highland Chavín Civilization: An Isotopic Perspective. *Am. Anthropol.* **1990,** *92,* 85–95.

Cross, H. Z.; Zuber, M. S. Prediction of Flowering Dates in Maize Based on Different Methods of Estimating Thermal Units. *Agron. J.* **1972,** *64,* 351–355.

Díaz-Ruiz, G.; Guyot, J. P.; Ruiz-Teran, F.; Morlon-Guyot, J.; Wacher, C. Microbial and Physiological Characterization of Weakly Amylolytic but Fast-Growing Lactic Acid Bacteria: A Functional Role in Supporting Microbial Diversity in Pozol, a Mexican Fermented Maize Beverage. *Appl. Environ. Microbiol.* **2003,** *69,* 4367–4374.

Dingerdissen, A. L.; Geige, H. H.; Lee, M.; Schechert, A.; Welz, H. G. Interval Mapping of Genes for Quantitative Resistance of Maize to *Setosphaeria turcica,* Cause of Northern Leaf Blight, in a Tropical Environment. *Mol. Breed.* **1996,** *2,* 143–156.

Eakin, H. Smallholder Maize Production and Climatic Risk: A Case Study from Mexico. *Climatic Change* **2000,** *45,* 19–36.

Farré, I.; Faci, J. M. Deficit Irrigation in Maize for Reducing Agricultural Water Use in a Mediterranean Environment. *Agric. Water Manag.* **2009,** *96,* 383–394. https://doi.org/10.1016/j.agwat.2008.07.002

Hastorf, C. A.; Johannessen, S. Pre-Hispanic Political Change and the Role of Maize in the Central Andes of Peru. *Am. Anthropol.* **1993,** *95,* 115–138.

Hertel, T.; Golub, A.; Jones, A.; O'Hare, M.; Plevin, R. J.; Kammen, D. M. Effects of US Maize Ethanol on Global Land Use and Greenhouse Gas Emissions: Estimating Market-Mediated Responses. *Bio Sci.* **2010,** *60,* 223–231. https://doi.org/10.1525/bio.2010.60.3.8

Hirasawa, T.; Hsiao, T. C. Some Characteristics of Reduced Leaf Photosynthesis at Midday in Maize Growing in the Field. *Field Crops Res.* **1999,** *62,* 53–62. https://doi.org/10.1016/S0378-4290 (99)00005-2

Hirel, B.; Bertin, P.; Quilleré, I.; Bourdoncle, W.; Attagnant, C.; Dellay, C.; Gouy, A.; Cadiou, S.; Retailliau, C.; Falque, M.; Gallais, A. Towards a Better Understanding of the Genetic and Physiological Basis for Nitrogen Use Efficiency in Maize. *Plant Physiol.* **2001,** *125,* 1258–1270. DOI: https://doi.org/10.1104/pp.125.3.1258.

Krause, V. M.; Tucker, K. L.; Kuhnlein, H. V.; Lopez-Palacios, C. Y.; Ruz, M.; Solomons, N. W. Rural-Urban Variation in Limed Maize Use and Tortilla consumption by Women in Guatemala. *Ecol. Food Nutr.* **1992,** *28,* 279–288.

Loyola-Vargas, V. M.; de Jimenez, E. S. Differential Role of Glutamate Dehydrogenase in Nitrogen Metabolism of Maize Tissues. *Plant Physiol.* **1984,** *76,* 536–540.

Otegui, M. E.; Andrade, F. H.; Suero, E. E. Growth, Water Use, and Kernel Abortion of Maize Subjected to Drought at Silking. *Field Crops Res.* **1995,** *40,* 87–94. https://doi.org/10.1016/0378-4290 (94)00093-

Pearsall, D. M. Maize is *Still* Ancient in Prehistoric Ecuador: The View from Real Alto, with Comments on Staller and Thompson. *J. Archaeol. Sci.* **2002,** *29,* 51–55.

Perales, H. R.; Benz, B. F.; Brush, S. B. Maize Diversity and Ethnolinguistic Diversity in Chiapas, Mexico. *PNAS*. **2005,** *102*, 949–954. https://doi.org/10.1073/pnas.0408701102

Pollak, M. L.; Corbett, J. D. Using GIS Datasets to Classify Maize-Growing Regions in Mexico and Central America. *Agron. J.* **1993,** *85*, 1133–1139.

Ranere A. J.; Piperno, D. R.; Holst, I.; Dickau, R.; Iriarte, J. The Cultural and Chronological Context of Early Holocene Maize and Squash Domestication in the Central Balsas River Valley, Mexico. *PNAS* **2009,** *106*, 5014–5018.

Ranum, P.; Peña-Rosas, J. P.; Garcia-Casal, M. N. Global Maize Productin, Utilization and Consumption. *Ann. New York Acad. Sci.* **2014,** *1312*, 101–112.

Seinfeld, D. M.; Nagy, von C.; Pohl, M. D. Determining Olmec Maize Use Through Bulk Stable Carbon Isotope Analysis. *J. Archaeol. Sci.* **2009,** *36*, 2560–2565. https://doi.org/10.1016/j.jas.2009.07.013

Soares, C.; Rodrigues, P.; Peterson, S. W.; Lima, N.; Venâncio, A. Three New Species of *Aspergillus* Section *Flavi* Isolated from Almonds and Maize in Portugal. *Mycologia* **2012,** *104*, 682–697.

Spedding, T. A.; Hamela, C.; Mehuysa, G. R.; Madramootoo, C. A. Soil Microbial Dynamics in Maize-Growing Soil Under Different Tillage and Residue Management Systems. *Soil Biol. Biochem.* **2004,** *36*, 499–512. https://doi.org/10.1016/j.soilbio.2003.10.026

Tojo Soler, C. M.; Sentelhas, P. C.; Hoogenboom, G. Application of the CSM-CERES-Maize Model for Planting Date Evaluation and Yield Forecasting for Maize Grown Off-Season in a Subtropical Environment. *Eur. J. Agron.* **2007,** *27*, 165–177. https://doi.org/10.1016/j.eja.2007.03.002

Willett, F. The Introduction of Maize into West Africa: An Assessment of Recent Evidence. *Africa.* **1962,** *32*, 1–13.

Zhang, J. Risk Assessment of Drought Disaster in the Maize-Growing Region of Songliao Plain, China. *Agric. Ecosyst. Environ.* **2004,** *102*, 133–153.

CHAPTER 2

World Maize Production

ABSTRACT

Maize is significant to the economy owing to its extensive uses. Maize is mainly used as feed for livestock, indicating the dependency of the livestock industry on maize production. It is also used to make a variety of food and nonfood products, such as corn meal, sweeteners, corn oil, starch and ethanol, which is used as a cleaner burning substitute to gasoline. In this chapter, the current maize production and advances made to meet future maize requirements are discussed.

2.1 MAIZE PRODUCTION AND COMMERCE

2.1.1 MAIZE PRODUCTION IN THE CORN BELT OF THE UNITED STATES

Maize crop production assessments are particularly precious. They influence the international trade and national economic policies and have large economic significance. Hayes and Decker (1996) investigated through the satellite data and estimated the maize production in the Corn Belt of the United States. They investigated this by usage of Vegetation Condition Index which has been obtained from NOAA/AVHRR satellite data. They have utilized the satellite data from 1985 to 1992 in a model and were able to elucidate for more than 50% of the variation in the normalized yields from almost 42 crop growing districts of the Corn Belt. They specified that these estimated results are more encouraging and this model would successfully give the operational estimates for regional maize production about 2 months before the harvest of maize in the Corn Belt.

Maize production in the United States, between 1910 and 1940 had no appreciable increase in yield per acre; however, the average yields from 1940 to 1955 had exhibited an increase of 50% and in the recent past decades, there was a further increase of 50% in yields. Aldrich and Leng (1965) in the book on Modern Corn Production have mentioned that the early drive to enhanced productivity is principally attributable to the development of hybrid varieties. However, the more recent advance in maize crop yields is chiefly the resultant of the improvements that occurred in the maize crop production techniques. Though the development of hybrid varieties has been comprehensively recognized, nevertheless it is only in recent times that ample accounts were published comprehensively of the innovations in methods of crop production. The results of recent research on various aspects of maize production were presented in a series of review articles in *Advances in Corn Production* (*F.C.A.* 20: 1422). The book is essentially a distillation of the data for the practicing farmer. They have produced a survey of modern maize production techniques in the Corn Belt States of America. Opening the discussion on maize with a general account of its growth and development from sowing to harvest, they later discussed the farming system, choice of variety, seedbed preparation, sowing, fertilizer requirements, water management, weed control, pests and diseases, feeding quality, prospects, etc. Although their approach throughout was in the form of a description of the practical methods for calibrating drills, sampling for soil and tissue tests, and identifying pests (including dishonest salesmen), recent research results are incorporated and discussed. Minimum cultivation techniques are flattering more and more popular and are being applicable in many areas in the Corn Belt. The commercial productions of maize are much oriented toward the earlier sowing, narrower drill rows, and higher plant populations. With changes in the design of farm machinery that has taken place and is continuing the authors believed that the erstwhile norm of 42-in. drill rows and 12–15 thousand plants per acre will soon be superseded by 30-in. drill rows and 24–28 thousand plants per acre. With the introduction of new technologies and research coupled with the innovative farming experiences on losses at harvest, combine fitted with a maize attachment may begin to replace the corn pickers. Further, previously there is the former practice of leaving the maize crop in the field until the grain moisture content was (22–25%) for the ears to be stored in cribs. This practice seems to be changing to harvesting the maize much earlier (30% grain moisture) and adopting the artificial drying methods of grain drying or storing it wet in sealed containers. It is evident that every facet of production, from seedbed preparation to harvesting and storing the crop as grain or silage has been the topic of intensive study in recent years. The book provides the maize farmer and advisory officer in the Corn Belt States of America with a fair and balanced appraisal of the present and potential effects of this research on farming practice.

2.1.2 ROOT ZONE WATER QUALITY MODEL (RZWQM) AND GENETIC COEFFICIENTS FOR THE CERES-MAIZE MODEL

In northeastern Colorado, maize production is mainly constrained due to frost. Frost free period in this region averages from 11 May to 27 September. Thus in this region, taking up the maize planting at an appropriate time in the growing season is very critical for attaining good hybrid maturity lengths and optimized yields. Many crop models were designed for determining in a locality the optimum windows of planting. Anapalli et al. (2005) calibrated the plant parameters of the RZWQM and genetic coefficients for the CERES-Maize model. They validated the performance of these models against experimental data of three corn hybrids that had exhibited variations in days to maturity and planted at three planting dates in 2 years at Akron, CO, under irrigation. Keeping in view the similar levels of accuracy in both these models they have calibrated them for prediction of leaf area index, soil water content, crop water use, and yield. The results have shown that there was a simulated drop in yield with late planting dates in both models though CERES-Maize simulated the yield from the latest planting date at a much accurate level for all three hybrids than did RZWQM (13% under predicted by CERES-Maize; 50% over predicted by RZWQM). They suggested through the use of a long-term Akron weather record that the most up-to-date planting dates for the short-, mid-, and long-season hybrids to have a 50% probability of attaining a breakeven yield under irrigation to be as 13 May, 20 May, and 6 May, respectively. Further, the results of long-term simulations have revealed that with delay in planting dates the hybrids with longer periods of maturity lengths would lose yields much faster rather than those with the short maturity length. Finally, their research highlights indicated that this information which has been produced by either RZWQM or CERES-Maize will be of much use in having decisions relating to both the planting and replanting in northeastern Colorado of corn hybrids varying in their maturity lengths.

2.1.3   THE HOUSEHOLDS' SOCIOECONOMIC CHARACTERISTICS AFFECTING MAIZE PRODUCTION

Maize crop in Sub Saharan African region occupies a vital position as the staple food in this area. Bunting (2015) examined the households' socioeconomic characteristics influencing maize production in Rukwa in the context of the market reforms carried out in Tanzania in the mid 1980s. It is one of Tanzania's most dependent maize producers. He explored specific issues of importance like farm size, education, and access to main inputs such as seeds, fertilizers, and agricultural extension services on the basis of data that has been gathered from three districts of Rukwa. The exploration results revealed that in most household's livelihoods maize crop had an important role to play, though its production levels were much low. An important factor concerned with bringing about an increase in yields is education. Some of the nonagricultural policies may also be vital to bring improvements in the productivity and welfare of farmers. In spite of its importance to household livelihoods, quite a lot of limitations were reported to impede its productivity including access to fertilizers, improved seeds, and other chemical inputs required for higher production and extension services.

2.1.4   ORDINARY LEAST SQUARE (OLS) METHOD

Hassan et al. (2014) used the secondary annual data of the periods of 1971–2010, and by using the OLS method analyzed the maize total factor productivity growth. They also used the Data Envelopment Analysis (DEA) that was based on Malmquist Index in Nigerian. By the OLS method they identified the factors affecting the maize productivity. Their analysis results have shown that total the factor productivity value was 1.004 for the 40-year period of maize production. The factor had a growth of 0.4%. Further, these results indicated that the country had recorded the total factor productivity growth of ≥ 1.00 that held at 43.6%, while 56.4% of the time studied the country had a decrease in maize total factor productivity growth. It has proved the inputs growth instead of an output growth. Apart from this, a regress of -3.5% was seen in the total factor productivity growth from 1971 to 1975, nevertheless from 1986 to 1990, there was 3.7% of maize productivity growth in the country. However, a 35.7% growth in maize productivity was experienced from 1991 to 1995 and a double digits productivity growth of 33.4% from 2006 to 2010. Research and development spending, net value of production, fertilizer price, and labor were recognized as the factors affecting the total factor productivity growth. Thus, they suggested that intensifying the scope of research and development, net value of production and labor use will help out to increase maize productivity in the country.

2.1.5 FACTORS INFLUENCING THE FORWARD PRICING METHODS DURING THE SEASON OF MAIZE PRODUCTION

Jordaan and Grové (2007) utilized a logistic regression for analyzing the factors affecting the choice of respondents in forward pricing methods during the season of maize production. Based on the results, it is assessed that minor levels of risk aversion and high levels of human capital are associated with the usage of forward pricing. They have employed the factor analysis for reducing the dimensionality of the personal reasons. These would be of help in interpreting the identical and similar factor that is behind the personal reasons of farmers being reluctant in using the forward pricing methods. They identified three factors viz deficiencies of capacity, distrust of the market, and the dreadful experiences. The outcomes from the factor analysis validate the finding that farmers require superior levels of human capital to use forward pricing methods. Further, they also don't consider that the method of forward pricing seems effective. They thus suggested that education be supposed to focus additionally on the practical application of option forward pricing methods and not solely on the profits of forward pricing methods utilization.

In the humid temperate regions of eastern Canada, even with the ease of use of modern hybrids and improved agronomic practices, there subsist outsized gaps between achievable yield of maize grown with recommended practices and producers' harvest yields. Subedi and Ma (2009) through a field experiment determined the key management yield limiting factors on rainfed maize grain production in Ottawa and given the recommendations of a few package of practices. The results revealed that weed infestation appeared to be a key factor in declining yields by 27–38% during the conditions of low diseases or insect's incidence. Further, it was observed that the absence of preplant N application (100 kg ha^{-1}) resulted in a yield decline of 10–22%. Similarly, though the low plant population density of (60,000 plants ha^{-1}) resulted in a decline in grain yields by 8–13% and an increase in plant population density of 90,000 plants ha^{-1} failed to bring about the improved yields. Though there was no effect of the withholding of P application on yield, deficiency, and lack of K resulted in a decline in yields by 13%, while Zn or Mn by 10% and 12%, respectively. Thus, they suggested that the absence of proper weed control, N fertilizer and plant population density are the key yield limiting factors of maize under rainfed conditions.

2.2   FUTURE REQUIREMENTS OF MAIZE PRODUCTION

2.2.1   CLIMATE CHANGE, CORN PRODUCTION

The changes in climate have a large impact on agriculture. It adds considerably to the challenges that aid in improving agricultural production and those aiming at the reduction of poverty and attainment of food security. Jones et al. (2003) through the usage of high resolution methods generated the daily characteristic data and generated a simulation model of the maize crop for assessing the effects of climate on production of maize in Africa and Latin America to 2055. The simulations data has indicated by and large a 10% decline in the maize production to 2055. This might be equal to losses of $2 billion per year. However, the combined results were found to hide massive inconsistencies such as areas that can be recognized where maize yields may vary to a large extent. Climate change straight away needs to be considered at the point of the household, so that poor and susceptible people reliant on agriculture can be suitably besieged in research and development actions whose objective is poverty lessening.

2.2.2   ASSESSMENT OF SUSCEPTIBILITY AND ADJUSTMENT OPTIONS OF THE MAIZE HYBRIDS

A chief maize producing province of China is Jilin. The maize production in this province has a key influence on the local as well as the nation's food security. Wang et al. (2011) developed the latest approach for assessing the susceptibility and adjustment options of the maize hybrids at Jilin with reference to the variabilities of climate. They modified a site-based biophysical model to a spatial grid-based application. In order to have complete information of the huge uncertainties of future predicted climates, using a combination of results obtained from 20 General Circulation Models (GCMs) and 6 results of a special report on emissions they developed an ensemble approach. The model results have indicated that in Western and Central regions of Jilin maize yields may exhibit a decline (15% rather than by 90%) by 2050 due to a reduction in the phenological growing season, where these hybrids are likely to exhibit shortened grain filling periods. However, in the eastern regions production of it was shown to exhibit an increase, even though its cultivation then was not taken up as a chief crop in the eastern region of Jilin. Further, an investigation carried out with augmented CO_2 level impacts on maize production has indicated that the predicted yield declines would be compensated with the elevated CO_2 levels. Conversely, additional field work and/or laboratory-based experiments are obligatory to corroborate the modeled CO_2 fertilization effects. They have realized two prospective adaptation approaches such as an improvement in the existing irrigation facilities and the introduction of the adaptive maize cultivars in line with the upcoming warming climate changes.

2.2.3   THE IMPACT OF CLIMATE CHANGE IN THE SEMI-HUMID AND SEMIARID, AGROCLIMATIC ZONES III–IV OF KENYA ON THE PRODUCTION OF MAIZE

Two GCMs: the Canadian Climate Center Model (CCCM) and the Geophysical Fluid Dynamics Laboratory (GFDL), as well as the CERES-Maize model were used by Mati (2000) to assess the effect of climate change in the semi-humid and semiarid, agroclimatic zones III–IV of Kenya on the maize production. They obtained from the meteorological stations of Kenya the lasting climate data, and maize data from six sites within the Kenya regions. They projected the climate scenarios up to 2030. Their investigation results of CCCM and GFLD have revealed that the regions in Kenya would experience a rise in temperature of 2·29 and 2·89°C, while rainfall remains to be unaffected. However, there are possibilities that rainfall patterns show shifts in their distribution. It is predicted that those regions which were currently are the short-rains season (October–January) are likely to the incidence of augmented rainfall, while those of the long rains season (April–July) will illustrate a decline in the rainfall. In zone III areas, the yields of maize are predicted to a reduction rather than those of zone IV areas which show increased yields. On the other hand, their predicted variations in yield changes were very meager, as these were below 500 kg ha^{-1}, in many regions of Kenya, except at the Homa Bay site. In consequence, of the predicted changes of maize yields in Kenya, they suggested that the cultivation of early maturing maize cultivars, adopting early planting methods will enable in counteracting the ill effects of climate change. Further, they also advocated for a shift of maize cultivation to eastern Kenya particularly for the short rain seasons.

2.2.4   CERES MAIZE MODEL TO PREDICT THE MAIZE PRODUCTION POTENTIAL IN CHINA

Xiong et al. (2007) predicted the maize production potential in China under irrigated and rainfed conditions by using a simulation CERES maize model under two climate change scenarios. They ran the model for 30 years of baseline climate and three time slices for the two climate change scenarios, without and with simulation of direct CO_2 fertilization effects. Their outcomes have revealed that climate change scenarios vary between the regions and years. At all times, under climate change scenarios of both A2 and B2, the maize production was predicted to be negatively affected and much impact may be seen in chief maize growing areas. Taking into consideration the effect of CO_2 fertilization, the model results revealed that rainfed maize yield production may exhibit an increase, while that of irrigated maize may exhibit a decline for the largest part of the time periods of both A2 and B2 scenarios.

2.2.5   STATISTICAL STUDIES OF RAINFED MAIZE YIELDS

Statistical studies of rainfed maize yields in the United States and other places have shown two apparent characteristics: a well-built negative yield response to buildup of temperatures above 30°C [or extreme degree days (EDD)], and a relatively weak response to seasonal rainfall. Lobell et al. (2013) have shown that the process-based Agricultural Production Systems Simulator (APSIM) is in a position to produce equally of these relationships in the Midwestern United States and provide insight into basic mechanisms. The principal special effects of EDD in APSIM are linked with enlarged vapor pressure deficit. It would contribute to water stress in two ways such as increasing demand for soil water to endure a given rate of carbon assimilation and decreasing the future supply of soil water by raising transpiration rates. Further, they suggested that daily water stress can effectively be computed by APSIM. It is computed as the ratio of water supply to demand. Their results revealed that during the critical month of July this ratio is three times additional responsive to 2°C warming than to a 20% precipitation decline. These findings put forward a reasonably negligible role for direct heat stress on reproductive organs at present temperatures in this region.

2.3 CONCLUSIONS

Maize is being cultivated for its high value processed products of great demand in the world. It has been observed that favorable climates lead to increased crop productivity. Water deficit and insects have been found to reduce productivity drastically. Research and development expenditure, net value of production, fertilizer price, and labor were identified as the factors affecting the total factor productivity growth. Inventive technologies have been developed to improve maize productivity.

KEYWORDS

- **maize**
- **yield**
- **global**
- **economy**
- **food and non-food products**

REFERENCES

Aldrich, S. R.; Leng, E. R. *Modern Corn Production;* CABI Abstract, **1965**; p 318.
Anapalli, S. S.; Ma, L.; Nielsen, D. C.; Vigil, F. M.; Ahuja, L. R. Simulating Planting Date Effects on Corn Production using RZWQM and CERES-Maize Models. *Agron. J.* **2005**, *97*, 58–71.

Bunting, E. S. Factors Influencing Maize Crop Production at Household Levels: A Case of Rukwa Region in the Southern Highlands of Tanzania. *Afr. J. Agric. Res.* **2015**, *10*, 1097–1106.

Hassan, Y.; Abdullah, A. M.; Ismail, M. M.; Mohamed, Z. A. Factors Influencing the Total Factor Productivity Growth of Maize Production in Nigeria. *IOSR J. Agric. Vet. Sci.* **2014**, *7*, 34–43.

Hayes, M. J.; Decker, W. L. Using NOAA AVHRR Data to Estimate Maize Production in the United States Corn Belt. *Int. J. Remote Sens.* **1996**, *17*(16), 3189–3200.

Jones, P. G.; Thornton, P. K. The Potential Impacts of Climate Change on Maize Production in Africa and Latin America in 2055. *Glob. Environ. Change* **2003**, *13*, 51–59.

Jordaan, H.; Grové, B. Factors Affecting Maize Producers Adoption of Forward Pricing in Price Risk Management: The Case of Vaalharts. *Agric. Econ. Res. Pol. Prac. South. Afr.* **2007**, *46*, 548–565.

Lobell, D. B.; Hammer, G. L.; McLean, G.; Messina, C.; Roberts, M. J.; Schlenker, W. The Critical Role of Extreme Heat for Maize Production in the United States. *Nat. Clim. Change* **2013**, *3*, 497–501.

Mati, B. M. The Influence of Climate Change on Maize Production in the Semi-Humid–Semi-Arid Areas of Kenya. *J. Arid Environ.* **2000**, *46*, 333–344.

Subedi, K. D.; Ma, B. L. Assessment of Some Major Yield-Limiting Factors on Maize Production in a Humid Temperate Environment. *Field Crops Res.* **2009**, *110*, 21–26.

Wang, M.; Li, Y.; Ye, W.; Bornman, J. F.; Yan, X. Effects of Climate Change on Maize Production and Potential Adaptation Measures: A Case Study in Jilin Province, China. *Climate Res.* **2011**, *46*, 223–242.

Xiong, W.; Matthews, R.; Holman, I.; Lin, E.; Xu, Y. Modelling China's Potential Maize Production at Regional Scale Under Climate Change. *Climate Change* **2007**, *85*, 433–451.

CHAPTER 3

Origin, Evolution, and Domestication of Maize

ABSTRACT

Maize is known from prehistoric times; it has been developed and culti-vated for millions of years. This chapter deals with the various domestica-tion theories and the spread of maize. The research advances made on the origin, evolution, and domestication of maize are also presented.

3.1 EVOLUTION OF CORN

Most of the plant and animal species were domesticated between 5000 and 10,000 years ago. During this period, several crops and animals were domesticated many times individually, which includes rice, common bean, millet, cotton, squash, cattle, sheep, and goats. Similarly, maize (*Zea mays* ssp. *mays*) has been deliberated to be the outcome of several independent domestications from its wild progenitor (teosinte) because of its amazing morphological and genetic diversity that occurs in it. Though the extraordinary diversity observed within maize was compatible with multiple domestications, it is also consistent with a single domestication and successive diversification. These two models can be differentiated by phylogenetic analyses, which include comprehensive samples of maize and its progenitor, teosinte. So, Matsuoka et al. (2002) undertook phylogenetic analyses of 264 individual plants and each plant was genotyped at 99 microsatellites. The findings specified that the maize was developed from a single domestication in southern Mexico around 9000 years ago. Further, they established that the maize types cultivated on Mexican highlands were the oldest surviving maize and spread from this region throughout the Americas along two major paths. The phylogenetic work was in agreement with a model based on the archaeological record representing that before the dispersal of maize in the lowlands it got diversified in the highlands of Mexico. Furthermore, they found only uncertain indications of post-domestication gene transfer from teosinte into maize.

The contemporary maize history commences at the dawn of human agriculture, approximately 10,000 years ago. In Mexico, farmers domesticated maize for the first time by just choosing its kernels (seeds) to plant. They observed that all plants were not the same. Some plants were taller than others, or may be some kernels tasted better or were easier to grind.

Then the plants with desired traits were collected by the farmers and were sowed for the following season's harvest. This method is known as selective breeding or artificial selection. Over time, maize cobs developed bigger, with more kernel rows, ultimately acquiring the form of modern maize.

Maize and its wild relatives, the teosintes, vary markedly in the morphology of their female inflorescences or ears. In spite of their different morphologies, many reports indicated that some teosinte varieties are cytologically identical with maize and able to form fully fertile hybrids with maize. Clark et al. (2004) performed molecular analysis of teosinte (*Z. mays* ssp. *parviglumis*) as the progenitor of maize. The morphological characters distinguishing maize and teosinte were analyzed. The outcome revealed that these traits are regulated by multiple genes and exhibits quantitative inheritance. However, these studies have also recognized some loci with large effects that appear to signify major advances during maize domestication. However, the major and minor effect genes, the polymorphisms within these genes controlling phenotypes, and in what way the individual and epistatic effects combination of these genes assisted the transformation of teosinte into maize is still unknown.

Maize (*Z. mays*) plant has a substantial economic importance as food-stuff and alternate energy source.

Researchers have come to an agreement that maize was domesticated from its wild relative teosinte (*Z. mays* spp. *parviglumis*) in Central America about 9000 years ago. In America, maize is known as corn, slightly confusing for the rest of the English-speaking world, where the seeds of any grain, such as barley, wheat, or rye are referred as "corn."

The maize has undergone substantial transformation during the domestication process. The seeds of wild teosinte are covered in hard shells and set on a spike with five to seven rows and when the grain is mature a spike shatters to disperse its seed. Modern maize has hundreds of uncovered kernels put together on a cob which is fully enclosed by husks and so cannot reproduce by itself. The morphological variation is among the highly divergent speciation known on the planet, and it is only current genetic studies that have established the association.

The initial domesticated maize cobs were from Guilá Naquitz Cave in Guerrero, Mexico, reported around 4280–4210 BC. The most primitive starch grains from domesticated maize have been observed in the Xihuatoxtla Shelter, in the Rio Balsas valley of Guerrero, around ~9000 BP.

Loss of seed shattering is a critical step during crop domestication. This study showed that a single gene, Shattering1 (Sh1) encodes a YABBY transcription factor that controls seed shattering in sorghum. Domesticated sorghums harbor three distinct mutations at the Sh1 locus. Variants at regulatory sites in the promoter and intronic regions bring about a low level of expression, a 2.2-kb deletion causes a truncated transcript that lacks exons 2 and 3, and a GT-to-GG splice-site variant in the intron 4 results in removal of the exon 4. The distributions of these nonshattering haplotypes among sorghum landraces suggest three independent origins. The function of the rice ortholog (OsSh1) was subsequently validated with a shattering-resistant mutant, and two maize orthologs (ZmSh1-1 and ZmSh1-5.1 + ZmSh1-5.2) were verified with a large mapping population. The study results indicated that Sh1 genes for seed shattering were under parallel selection during sorghum, rice, and maize domestication (Lin et al., 2012).

Domestication and plant breeding are continuing 10,000-year-old evolutionary researches that have drastically transformed wild species to meet human requirements. Maize has undergone a particularly remarkable transformation. Researchers worked for decades to find the genes that lie behind maize evolution, but these works have been limited in scope. In this chapter, a broad assessment of the evolution of modern maize was reported based on the genome-wide resequencing of 75 wild, landrace, and superior maize lines. They found evidences of recovery of diversity after domestication, likely introgression from wild relatives, and proof for stronger selection during domestication than improvement. They identified a number of genes with stronger signals of selection than those formerly shown to bring about major morphological variations. Lastly, through transcriptome-wide analysis of gene expression, they found evidence both reliable with removal of cis-acting variation during maize domestication and improvement and suggestive of modern breeding having improved dominance in expression while targeting highly expressed genes (Hufford et al., 2012).

The maize landraces developed by pre-Columbian cultivators contain extraordinary morphological and genetic diversity. To elucidate this high level of diversity in maize, a number of authors have suggested that maize landraces were the products of several independent domestications from their wild relative (teosinte). They presented phylogenetic analyses based on 264 individual plants, each genotyped at 99 microsatellites, which challenges the multiple-origins hypothesis. Instead, the findings indicated that all maize developed from a single domestication in southern Mexico about 9000 years ago. Further, the analyses indicated that the oldest enduring maize types are those of the Mexican highlands with maize dispersing from this region over the Americas along two main paths. The phylogenetic work is in harmony with a model based on the archaeological record signifying that maize diversified in the highlands of Mexico before spreading to the lowlands. Further, they found only modest evidence for post-domestication gene flow from teosinte into maize (Matsuoka et al., 2002).

Ten thousand years ago human societies around the globe started to transform from hunting and gathering to agriculture. By 4000 years ago, ancient peoples had accomplished the domestication of all major crop species upon which human continued existence is dependent, comprising rice, wheat, and maize. Current research has initiated to make known the genes accountable for this agricultural revolution. The list of genes to date tentatively suggested that diverse plant developmental pathways were the targets of Neolithic "genetic tinkering," and are now closer to understanding how plant development was redirected to encounter the needs of a hungry world (Doebley et al., 2006).

All major crop plants were domesticated during a brief period in human history around 10,000 years ago. During this period, early agriculturalists selected seeds of preferred forms and culled out seeds of undesirable types to produce each successive generation. Accordingly, favored alleles at genes controlling traits of interest increased in frequency, finally reaching fixation. When selection is strong, domestication has the potential to considerably reduce genetic diversity in a crop. To comprehend the influence of selection during maize domestication, nucleotide polymorphism in teosinte branched1 (tb1), a gene involved in maize evolution was examined. Wang et al. (1999) showed that the effects of selection were restricted to the gene's regulatory region and cannot be identified in the protein-coding region. Though selection was apparently strong, high rates of recombination and a prolonged domestication period possibly limited its effects. The study results aid to elucidate why maize is such a variable crop. Further, they suggested that maize domestication required hundreds of years, and strengthen earlier proof that maize was domesticated from Balsas's teosinte of southwestern Mexico.

The most significant step in maize (*Z. mays* ssp. *mays*) domestication was the release of the kernel from the hardened, protective casing that envelops the kernel in the maize progenitor, teosinte. This evolutionary step exposed the kernel on the surface of the ear, such that it could readily be used by humans as a food source. Wang et al. (2005) demonstrate that this significant event in maize domestication is regulated by a single gene (*teosinte glume* architecture or *tga1*), belonging to the SBP-domain family of transcriptional regulators. The factor regulating the phenotypic variance between maize and teosinte maps to a 1-kb region, within which maize and teosinte show only seven predetermined variations in their DNA sequences. One of these variations encodes a nonconservative amino acid substitution and may affect protein function, and the other six variations potentially affect gene regulation. Molecular evolution analyses exhibit that this region was the target of selection during maize domestication. The findings demonstrated that modest genetic variations in single genes can induce remarkable variations in phenotype during domestication and evolution.

The last two decades have viewed significant progresses in our understanding of maize domestication, thanks partly to the contributions of genetic data. Genetic studies have made available strong proof that maize was domesticated from Balsas teosinte (*Z. mays* ssp. *parviglumis*), a wild relative that is endemic to the mid- to lowland regions of southwestern Mexico. An interesting paradox remains; however, maize cultivars that are most closely associated with Balsas teosinte are located mainly in the Mexican highlands where subspecies parviglumis do not grow. Genetic data thus point to the initial diffusion of domesticated maize from the highlands rather than from the region of initial domestication. The latest archeological evidence for early lowland cultivation has been consistent with the genetics of domestication, leaving the issue of the ancestral position of highland maize unresolved. Here, a new SNP dataset scored in a large number of accessions of both teosinte and maize was used to take a second look at the geography of the most primitive cultivated maize. It was observed that gene flow between maize and its wild relatives meaningfully impact our inference of geographic origins. By examining differentiation from inferred ancestral gene frequencies, they got results that are fully in agreement with present ecological, archeological, and genetic data relating to the geography of early maize cultivation (Van Heerwaarden et al., 2011).

The natural history of maize commenced nine thousand years ago when Mexican farmers began collecting the seeds of the wild grass, teosinte. Valuable as a food source, maize permeated Mexican culture and religion. Its domestication eventually led to its acceptance as a model organism, assisted in large part by its large chromosomes, ease of pollination, and rising agricultural significance. Genome comparisons between varieties of maize, teosinte, and other grasses are the foundation to find the genes accountable for the domestication of modern maize and are also presenting ideas for the breeding of more hardy varieties (Hake et al., 2015).

How domestication bottlenecks and artificial selection shaped the amount and distribution of genetic variation in the genomes of modern crops is not well known. In this study, diversity at 462 simple sequence repeats (SSRs) or microsatellites spread throughout the maize genome was analyzed and the diversity seen at these SSRs in maize was compared with that observed in its wild progenitor, teosinte. The results unveiled a modest shortage of diversity in the maize genome comparative to teosinte. The relative shortage of diversity is less for SSRs with dinucleotide repeat motifs than for SSRs with repeat motifs of more than two nucleotides, signifying that the former with their higher mutation rate have partly recovered from the domestication bottleneck. The association between SSR diversity and proximity to QTL for domestication traits was analyzed and association was not found between these factors. But, a weak, though significant, spatial correlation was observed for diversity statistics among SSRs within 2 cM of one another, indicating that SSR diversity is weakly patterned across the genome. Twenty-four of 462 SSRs (5%) reveal some indication of positive selection in maize under multiple tests. On the whole, the bottleneck effect can largely elucidate the pattern of genetic diversity at maize SSRs with a smaller effect from selection (Vigouroux et al., 2005).

Queries that still surround the origin and early dispersals of maize (*Z. mays* L.) bring about a large part from the lack of information on its primary history from the Balsas River Valley of tropical southwestern Mexico, where its wild ancestor is native. Here, starch grain and phytolith data were reported from the Xihuatoxtla shelter, located in the Central Balsas Valley, that indicated that maize was present by 8700 calendrical years ago (cal. B.P.). Phytolith data also showed a primary preceramic existence of a domesticated species of squash, probably *Cucurbita argyrosperma*. The starch and phytolith data also allowed an assessment of current assumptions about how early maize was utilized and make available evidence of the tempo and timing of human selection pressure on two major domestication genes in *Zea* and *Cucurbita*. The study confirmed an early Holocene chronology for maize domestication that has been formerly shown by archaeological and paleoecological phytolith, starch grain, and pollen data from south of Mexico, and reshift the emphasis back to an origin in the seasonal tropical forest rather than in the semiarid highlands (Piperno et al., 2009).

The domestication of crop plants has often involved an increase in apical dominance (the concentration of resources in the main stem of the plant and a corresponding suppression of axillary branches). A prominent example of this phenomenon is observed in maize (*Z. mays* spp. *mays*), which shows a profound increase in apical dominance compared with its likely wild ancestor, teosinte (*Z. mays* ssp. *parviglumis*). Earlier research has recognized the tb1 gene as a key contributor to this evolutionary trans-formation in maize. Doebley et al. (1997) cloned tb1 by transposon tagging and showed that it encodes a protein with homology to the cycloidea gene of snapdragon. The pattern of tb1 expression and the morphology of tb1 mutant plants suggested that tb1 acts both to repress the growth of axillary organs and to facilitate the formation of female inflorescences. The maize allele of tb1 is expressed at twice the level of the teosinte allele, signifying that gene controlling variations lie behind the evolutionary divergence of maize from teosinte.

Domesticated maize and its wild ancestor (teosinte) differ markedly in morphology and provide an opportunity to assess the association between strong selection and diversity in a main crop species. The tb1 gene mainly regulates the increase in apical dominance in maize comparative to teosinte, and a region of the tb1 locus 5′ to the transcript sequence was a target of selection during maize domestication. To better characterize the influence of selection at a key "domestication" locus, Clark et al. (2004) sequenced the upstream tb1 genomic region and systematically sampled nucleotide diversity for sites located as far as 163 kb upstream to tb1. Their studies defined a selective sweep of about 60–90 kb 5' to the tb1-transcribed sequence. The selected region harbors a combination of unique sequences and large repetitive elements, but it holds no predicted genes. Diversity at the nearest 5' gene to tb1 is characteristic of that for neutral maize loci, representing that selection at tb1 had the least effect on the adjacent chromosomal region. Further, results showed low intergenic linkage disequilibrium in the region and suggested that selection had a negligible role in shaping the pattern of linkage disequilibrium observed. In brief, the study increases the likelihood that maize-like tb1 haplotypes are present in extant teosinte populations, and their results also suggested a model of tb1 gene regulation that varies from traditional opinions of how plant gene expression is regulated.

Accelerator mass spectrometry age determinations of maize cobs (*Z. mays* L.) from Guilá Naquitz Cave in Oaxaca, Mexico, produced dates of 5400 carbon-14 years before the present (about 6250 calendar years ago), producing those cobs the oldest in the Americas. Macrofossils and phytoliths traits of wild and domesticated *Zea* fruits are absent from older strata from the site, although *Zea* pollen has previously been identified from those levels. These results, along with the modern geographical distribution of wild *Z. mays*, suggested that the cultural practices that led to *Zea* domestication may be taken place somewhere else in Mexico. Guilá Naquitz Cave has now yielded the primary macrofossil evidence for the domestication of two major American crop plants, squash (*Cucurbita pepo*) and maize (Piperno and Flannery, 2001).

Although man does not produce variability and can not even avert it, he can select, preserve, and gather the variations given to him by the hand of nature almost in any way which he chooses; and thus he can certainly produce a great result. Wild on a Mexican hillside grows teosinte, its scanty ear comprising only two entwined rows of small, well-armored kernels. This unassuming grass might easily have been ignored, were it not for the hand of nature that beckoned with abundant variation, a gift not lost on early agriculturists. Within the last 10,000 years, early Native Americans were able to transform teosinte into a plant whose ear, bursting with row upon row of exposed kernels, feeds the world over. It was a transformation so striking and so complex that some would not believe it probable, leading to years of competing theory and intense debate. But as Darwin himself accepted, when human desires collide with the diversity of nature, the result can be great indeed. Although debate still remains over the origin of maize, the molecular revolution of the last decade has presented convincing evidence in support of teosinte as the progenitor of modern maize. This chapter reviews that evidence in light of some different domestication assumptions. Further, it discusses the rich genetic diversity at the source of such an extraordinary morphological conversion and examines how human selection has influenced this diversity, both at individual loci and for a whole metabolic pathway (Buckler and Stevens, 2016).

Artificial selection results in phenotypic evolution. Maize (*Z. mays* L. ssp. *mays*) was domesticated from its wild progenitor teosinte (*Z. mays* ssp. *parviglumis*) through a specific domestication event in southern Mexico between 6000 and 9000 years ago. This domestication event leads to original maize landrace varieties. The landraces make available the genetic material for modern plant breeders to select improved varieties and inbred lines by improving traits regulating agricultural productivity and performance. Artificial selection during domestication and crop improvement involved the selection of specific alleles at genes regulating main morphological and agronomic traits, resulting in decreased genetic diversity comparative to unselected genes. In this review, research on the identification and characterization by population genetics methods of genes affected by artificial selection in maize was summarized. Analysis of DNA sequence diversity at a large number of genes in a sample of teosintes and maize inbred lines showed that about 2% of maize genes reveal evidence of artificial selection. The remaining genes provide evidence of a population bottleneck related to domestication and crop improvement. In a further study to efficiently find selected genes, the genes with zero sequence diversity in maize inbreds were selected as potential targets of selection and sequenced in diverse maize landraces and teosintes, resulting in approximately half of candidate genes showing evidence for artificial selection. Extended gene sequencing demonstrated a low false-positive rate in the approach. The selected genes have functions consistent with agronomic selection for plant growth, nutritional quality, and maturity. Large-scale screening for artificial selection permits the detection of genes of potential agronomic significance even when gene function and the phenotype of interest are unknown. These methods should also be appropriate for other domesticated species if precise demographic conditions during domestication exist (Yamasaki et al., 2007).

Domestication is a controllable system for subsequent evolutionary change. Under domestication, wild populations respond to unstable selective pressures, resulting in adaptation to the new ecological niche of cultivation. Because of the vital role of domesticated crops in human nutrition and agriculture, the pedigree and selection pressures transforming a wild plant into a domesticated plant have been comprehensively studied. In *Z. mays*, morphological, genetic, and genomic studies have explained how a wild plant, the teosinte *Z. mays* subsp. *parviglumis*, was transformed into the domesticate *Z. mays* subsp. *mays*. Five key morphological variances differentiate these two subspecies, and careful genetic dissection has identified the molecular changes accountable for several of these traits. However, maize domestication was a result of more than just five genes, and regions all over the genome contribute. The effects of these additional regions are depending on genetic background, both the interactions between alleles of a single gene and among alleles of the multiple genes that modulate phenotypes. Important genetic interactions contain dominance relationships, epistatic interactions, and pleiotropic constraints, including how these variants are connected in gene networks. Here, Stitzer and Ross-Ibarra (2018) reviewed the role of gene interactions in creating the remarkable phenotypic evolution noticed in the transition from teosinte to maize.

The origin of maize (*Z. mays*) in the US Southwest remains controversial, with contradictory archaeological data supporting either coastal or highland routes of maize diffusion into the United States. Additionally, the genetics of adaptation to the new environmental and cultural background of the Southwest is mostly uncharacterized. To address these issues, Da Fonseca et al. (2015) compared nuclear DNA from 32 archaeological maize samples spanning 6000 years of evolution to modern landraces. They found that the primary diffusion of maize into the Southwest approximately 4000 years ago is probable to have taken place along a highland route, followed by gene flow from lowland coastal maize starting at least 2000 years ago. The population genetic studies also permitted the differential selection during domestication for adaptation to the climatic and cultural environment of the Southwest, detecting adaptation loci pertinent to drought tolerance and sugar content.

Phenotypic variation has been influenced by humans during crop domestication, which occurred mainly between 3000 and 10,000 years ago in the different centers of origin across the world. The process of domestication had great consequences on crops, where the domestication has moderately decreased genetic diversity comparative to the wild ancestor across the genome, and severely reduced diversity for genes targeted by domestication. The question that remains is whether a reduction in genetic diversity has influenced crop production today. A case study in maize (*Z. mays*) established the application of understanding associations between genetic diversity and phenotypic diversity in the wild ancestor and the domesticated. As an outcrossing species, maize has remarkable genetic variation. The complementary combination of genome-wide association mapping (GWAS) methods, large HapMap data sets, and germplasm resources is leading to significant findings of the association between genetic diversity and phenotypic variation and the effect of domestication on trait variation (Flint-Garcia, 2013).

Domestication research has mainly focused on the detection of morphological and genetic variances between extant populations of crops and their wild relatives. Much attention has not been paid to the probable effects of the environment in spite of considerable known variations in climate from the time of domestication to the modern day. Current research, in which maize and teosinte (i.e., wild maize) were subjected to environments similar to the time of domestication, resulted in a plastic induction of domesticated phenotypes in teosinte and little response to the environment in maize. These findings indicated that early agriculturalists might have selected for genetic mechanisms that cemented domestication phenotypes primarily induced by a plastic response of teosinte to the environment, a process known as genetic assimilation. To get a better understanding of this phenomenon and the probable role of the environment in maize domestication, Lorant et al. (2017) studied differential gene expression in maize (*Z. mays* ssp. *mays*) and teosinte (*Z. mays* ssp. *parviglumis*) between past and present conditions. They found a gene set of over 2000 loci exhibiting a variation in expression across environmental conditions in teosinte and invariance in maize. On the whole, they observed both greater plasticity in gene expression and more significant rewiring of expression networks in teosinte across environments when compared to maize. While these findings suggested genetic assimilation played at least some role in domestication, genes showing expression patterns consistent with assimilation are not significantly enriched for formerly recognized domestication candidates, specifying assimilation did not have a genome-wide effect.

Over the past decade, growing interest in the recovery and identification of plant microfossil remnants from archaeological sites located in lowland South America has considerably improved knowledge of pre-Columbian plant domestication and crop plant dispersals in tropical forests and other regions. Along the Andean mountain chain, however, the chronology and path of plant domestication are still not understood well for both vital indigenous staple crops such as the potato (*Solanum* sp.) and others exogenous to the region, for example, maize (*Z. mays*). Perry et al. (2006) reported the studies of plant micro remains from a late preceramic house (3,431 ± 45 to 3745 ± 65 ^{14}C BP or ~3600 to 4000 calibrated years BP) in the highland southern Peruvian site of Waynuna. Their findings extend the record of maize by at least a millennium in the southern Andes, show on-site processing of maize into flour, make available direct evidence for the purposeful movement of plant foods by humans from the tropical forest to the highlands, and confirm the potential of plant microfossil analysis in knowing ancient plant use and migration in this region.

Maize was initially domesticated in a restricted valley in south-central Mexico. It was diffused throughout the Americas over thousands of years, and following the discovery of the New World by Columbus, was introduced into Europe. Trade and colonization introduced it further into all parts of the world to which it could adapt. Repeated introductions, local selection and adaptation, a highly diverse gene pool and out-crossing nature, and global trade in maize caused problems in understanding exactly where the diversity of many of the local maize landraces originated. This is mostly true in Africa and Asia, where historical accounts are limited or contradictory. Knowledge of post-domestication movements of maize across the world would help in germplasm conservation and plant breeding efforts. To this end, Mir et al. (2013) utilized SSR markers to genotype multiple individuals from hundreds of representative landraces from around the world. Employing a multidisciplinary approach combining genetic, linguistic, and historical data, they reconstructed probable patterns of maize diffusion all through the world from American "contribution" centers, which reflect the origins of maize worldwide. These findings provided an insight into the introduction of maize into Africa and Asia. By giving a first globally comprehensive genetic characterization of landraces using markers suitable to this evolutionary time frame, they explored the post-domestication evolutionary history of maize and highlighted original diversity sources that may be exploited for plant improvement in diverse regions of the world.

The complex evolutionary history of maize (*Z. mays* L. ssp. *mays*) has been explained with genomic-level data from modern landraces and wild teosinte grasses, boosting archaeological findings that suggested domestication has taken place between 10,000 and 6250 years ago in southern Mexico. Maize evolved rapidly under human selection, resulting in noticeable phenotypic transformations, as well as adaptations to diverse environments. Still, many doubts about the domestication process remain unanswered because modern specimens do not represent the full range of past diversity due to desertion of unproductive lineages, genetic drift, on-going natural selection, and modern breeding activity. To better understand the history and disperse of maize, Ramos-Madrigal et al. (2016) characterized the draft genome of a 5310-year-old archaeological cob mined in the Tehuacan Valley of Mexico. They compared this ancient sample against a reference panel of modern landraces and teosinte grasses using D statistics, model-based clustering algorithms, and multidimensional scaling analyses, representing the specimen obtained from the same source population that gave rise to modern maize. They found that 5310 years ago, maize in the Tehuacan Valley was genetically closer to modern maize than to its wild counterpart. But, many genes linked with important domestication traits existed in the ancestral state, sharply contrasting with the ubiquity of resultant alleles in living landraces. These results suggested much of the evolution during domestication may have been gradual and boost further paleogenomic research to deal with provoking questions about the world's most-produced cereal.

The domestication of maize (*Z. mays* sp. *mays*) from its wild progenitors signifies a prospect to study the timing and genetic basis of morphological divergence resultant from artificial selection on target genes. Hufford et al. (2007) compared sequence diversity of 30 candidate selected and 15 reference loci between the three populations of wild teosintes, maize landraces, and maize inbred lines. They inferred almost an equal ratio of genes chosen during early domestication and genes chosen during current crop breeding. By means of an extended dataset of 48 candidate selected and 658 neutral reference loci, they verified the assumption that candidate selected genes in maize are more probable to have transcriptional functions than neutral reference genes, but there was no overrepresentation of regulatory genes in the chosen gene dataset. Electronic northern analysis revealed that candidate genes are significantly overexpressed in the maize ear comparative to vegetative tissues such as maize shoot, leaf, and root tissue. The maize ear experienced remarkable morphological change upon domestication and has been a continuing target of selection for maize yield. So, they put forward that genes targeted by selection are more likely to be expressed in tissues that underwent high levels of morphological variance during domestication and crop improvement.

Strong directional selection has taken place during the domestication of maize from its wild ancestor teosinte, reducing its genetic diversity, mainly at genes controlling traits related to domestication. However, variability for some domestication-related traits is preserved in maize. The genetic basis of this could be sequence variation at the same key genes regulating differentiation of maize–teosinte, different loci with large effects, or polygenic background variation. Earlier studies allow annotation of maize genome regions linked with the major variances between maize and teosinte or that reveal population genetic signals of selection during either domestication or post-domestication improvement. Genome-wide association studies and genetic variance partitioning analyses were carried out in two dissimilar maize inbred line panels to compare the phenotypic effects and variances of sequence polymorphisms in regions involved in domestication and improvement to the rest of the genome. Additive polygenic models elucidated most of the genotypic variation for domestication-related traits; no large effect loci were identified for any trait. Maximum trait variance was linked with background genomic regions lacking earlier evidence for contribution in domestication. Improvement sweep regions were linked with more trait variation than expected based on the proportion of the genome they represent. Selection during domestication removed large effect genetic variants that would revert maize toward a teosinte type. Small effect polygenic variants were accountable for most of the standing variation for domestication-related traits in maize (Xue et al., 2016).

Modern maize was domesticated from *Z. mays parviglumis*, a teosinte, about 9000 years ago in Mexico. Genes assumed to have been selected upon during the domestication of crops are normally known as domestication loci. The ramosa1 (ra1) gene encodes a putative transcription factor that regulates branching architecture in the maize tassel and ear. Earlier work established lessened nucleotide diversity in a segment of the ra1 gene in a study of modern maize inbreds, showing that positive selection happened at some point in time since maize diverged from its common ancestor with the sister species *Tripsacum dactyloides* and encouraging the assumption that ra1 may be a domestication gene. To test this assumption, Sigmon and Vollbrecht (2010) studied ear phenotypes resultant from minor changes in ra1 activity and sampled nucleotide diversity of ra1 across the phylogenetic spectrum between tripsacum and maize, containing a wide panel of teosintes and unimproved maize landraces. Weak mutant alleles of ra1 exhibited subtle effects in the ear, with crooked rows of kernels because of the occasional formation of extra spikelets, associating a probable, selected trait with subtle variations in gene activity. Nucleotide diversity was significantly reduced for maize landraces but not for teosintes, and statistical tests suggested directional selection on ra1 consistent with the assumption that ra1 is a domestication locus. In maize landraces, a noncoding 3'-segment had almost no genetic diversity and 5'-flanking diversity was significantly reduced, indicating that a regulatory element might have been a target of selection.

3.2   MAIZE DOMESTICATION THEORIES

Two main theories have been proposed by scientists regarding the growth of maize.

The teosinte model deliberates that maize developed directly from teosinte through genetic mutation in the lowlands of Guatemala. The hybrid origin model argues that maize developed as a hybrid of diploid perennial teosinte and early-stage domesticated maize in the highlands of Mexico. Eubanks has proposed a parallel development of maize within the Mesoamerican interaction domain between lowland and highland.

In recent times, evidence of starch grain has been located in Panama signifying the usage of maize there around 7800 to 7000 BP, and the discovery of wild teosinte cultivation in the Balsas river region of Mexico supported that model.

The domesticated maize starch granules in occupation levels courted to the Paleoindian period, more than 8990 BP were discovered in the Xihuatoxtla rock shelter of the Balsas river region. This indicates that hunter-gatherers might have domesticated maize for thousands of years before it developed as a staple of people's diets.

The maize genome is the outcome of a whole-genome duplication that generated two duplicate genomes. Each duplicated genome is orthologous to the complete sorghum genome that has been reduced by fractionation. The evolution of inflorescence between the two species is found to be analogous to the fractionation process that is the perfect flowers of sorghum's single inflorescence fractionating into imperfect male and female flowers on distinct inflorescences in maize. Left panel: maize male flower; middle panel: sorghum flower; right panel: ear of maize with female tassels at the top.

Evolution is an uncertain process. From one generation to the next, the millions or billions of base pairs that constitute an organism's genetic material experience negligible but occasionally noteworthy variations. The addition or deletion of base pairs that create new combinations may make the plants better than the earlier viz., its parents. The path to biodiversity is littered with the brief recollection of genetic remixes that made their short-lived possessors incompetent for existence.

3.3   THE SPREAD OF MAIZE

Through the dispersal of seeds along trade networks rather than migration of people the maize eventually spread out from Mexico. Maize was cultivated in the southwestern United States around 3200 years ago and in the eastern United States around 2100 years ago. Maize became well established into the Canadian Shield by 700 AD.

DNA studies suggested that determined selection for desired characters sustained all over this period that lead to the development of wide variety of species today. For instance, 35 diverse races of maize have been recognized in pre-Columbian Peru, which includes popcorns, flint varieties, and varieties for specific purposes, for example, chicha beer, textile dyes, and flour.

During their evolutionary history plant species, such as maize, wheat, and so on, experienced a process that made them polyploid, namely, increased the number of chromosome pairs, instead of just the single pair possessed by the majority of animals. These additional chromosome pairs are likely to make them better and more productive from the human viewpoint. Further, they make available more resources for evolution and allow them to carry out key changes in genetic makeup without losing basic function.

Four years ago, researchers found that the two sets of chromosomes in *Arabidopsis* were differentially transformed, or fractionated, with more of the alterations accumulating on one set than the other. It was observed that this discrimination in two chromosomes is adaptive and it maintains the original pair intact, offering a safety net and the other homeolog (matched chromosome) progressively sheds and recombines genetic material, ridding excess and making new combinations.

Simultaneously, plant biologists Woodhouse et al. (2010) from the University of California Berkeley undertook a comparison of two newly sequenced grasses—sorghum and B73 inbred maize. They compared 37 stretches of sorghum DNA containing 2943 shared genes which represent "before evolution" proxy picture with the "after evolution" present maize genome. It was observed that the maize reserved about 43% of the genes, with an uneven share of the retained genes encoding transcription factors. When they examined genetic material stretches still present in sorghum but lacking or transformed in maize, they found that like *Arabidopsis*—corn underwent biased fractionation, with alteration taking place 2.3 times more often on one of the homeologs than the other. The majorities of these alterations were gene deletions and happened on a gene-by-gene basis rather than in clusters. The homeolog with the majority of the variations was expected to be missing transposons and other dispensible DNA making up the bulk of the genome. The scientists revealed that the DNA in between genes was more likely to be differentially fractionated than genes themselves.

Later, the researchers compared the maize gene remnants with sorghum and rice. They found that the deleted sequences frequently have short, identical sequences on both sides of them in the secondary progenitor. This put forward that the mechanism for fractionation here is mainly "illegitimate recombination"—a form of sequence deletion in which adjacent identical sequences line up, making a loop out of the base pairs between them that finally pinches off. Then they analyzed whether this event was associated with the existence of methylated domains. Though it was not, it is still probable that another form of epigenetic mark, such as histone modification, may well be at play. The researchers noted that, from an evolutionary viewpoint, biased fractionation makes lots of sense. A mutation that deactivates one homeolog automatically sets strong selective pressure against a debilitating mutation on the other. Alternatively, constant mutations on the previously inactivated chromosome can not do any more harm. Finally, the selection (natural or otherwise) forces remove the extra DNA from the tetraploid and creates new combinations and associations to face its everchanging environment and, in maize's case, meet evergrowing demands to make available food and fuel to an increasing human population.

3.4 CONCLUSIONS

Many studies have been conducted on the origin, evolution, and domestication of maize in different countries based on its archeological, phylogeny, cytological, and molecular levels. Several theories were put forth by different authors on these aspects. Various studies deliberated the origin of maize, the process of evolution, polyploidization, and hybridization that contributed to the improvement of modern maize.

KEYWORDS

- maize
- origin
- evolution
- research advances

REFERENCES

Buckler, E. S.; Stevens, N. M. Maize Origins, Domestication, and Selection. In *Darwins Harvest*, 2016 ed.; Motley T.; Columbia University Press: New York Chichester, West Sussex, **2016,** Chapter 4.

Clark, R. M.; Linton, E.; Messing, J.; Doebley, J. F. Pattern of Diversity in the Genomic Region Near the Maize Domestication Gene tb1. *Proc. Natl. Acad. Sci. U.S.A.* **2004,** *101,* 700–707.

Da Fonseca, R. R.; Smith, B. D.; Wales, N.; Cappellini, E.; Skoglund, P.; Fumagalli, M.; Samaniego, J. A.; Carøe, C.; Ávila-Arcos, M. C.; Hufnagel, D. E.; Korneliussen, T. S.; Vieira, F. G.; Jakobsson, M.; Arriaza, B.; Willerslev, E.; Nielsen, R.; Hufford, M. B.; Albrechtsen, A.; Ross-Ibarra, J.; Gilbert, M. T. P. The Origin and Evolution of Maize in the Southwestern United States. *Nat. Plants* **2015,** *1,* 14003.

Doebley, J.; Adrian, S.; Hubbard, L. The Evolution of Apical Dominance in Maize. *Nature* **1997,** *386,* 485–488.

Doebley, J. F.; Gaut, B. S.; Smith, B. D. The Molecular Genetics of Crop Domestication. *Cell* **2006,** *127,* 1309–1321.

Doebley, J. The Genetics of Maize Evolution. *Ann. Rev. Genet.* **2004,** *38,* 37–59.

Flint-Garcia, S. A. Genetics and Consequences of Crop Domestication. *J. Agric. Food Chem.* **2013,** *61,* 8267–8276.

Hake, S.; Ross-Ibarra, J. Genetic, Evolutionary and Plant Breeding Insights from the Domestication of Maize. *eLife* **2015,** *4,* 23–31.

Hoff, M. A Window on Maize Evolution. *PLoS Biol.* **2010,** *8* (6), e1000411.

Hufford, M. B.; Xun, X.; Heerwaarden, J. V.; Pyhäjärvi, T.; Chia, J. M.; Cartwright, R. A.; Elshire, R. J.; Glaubitz, J. C.; Guill, K. E.; Kaeppler, S. M.; Lai, J.; Morrell, P. L.; Shannon, L. M.; Song, C.; Springer, N. M.; Swanson-Wagner, R. A.; Tiffin, P.; Wang, J.; Gengyun, Z.; Doebley, J.; McMullen, M. D.; Ware, D.; Buckler, E. S.; Yang, S.; Ross-Ibarra, J. Comparative Population Genomics of Maize Domestication and Improvement. *Nat. Genet.* **2012,** *44,* 808–811.

Hufford, K. M.; Canaran, P.; Ware, D. H.; McMullen, M.; D.; Gaut, B. S. Patterns of Selection and Tissue-Specific Expression among Maize Domestication and Crop Improvement loci. *Plant Physiol.* **2007**, *144*, 1642–1653.

Lin, Z.; Li, X.; Shannon, L. M.; Yeh, C.-T.; Wang, M. L.; Bai, G.; Peng, Z.; Li, J.; Trick, H. N.; Clemente, T. E.; Doebley, J.; Schnable, P. S.; Tuinstra, M. R.; Tesso, T. T.; White, F.; Yu, J. Parallel Domestication of the Shattering1 genes in Cereals. *Nat. Genet.* **2012**, *44*, 720–724.

Lorant, A.; Pedersen, S.; Holst, I.; Hufford, M. B.;Winter, K.; Piperno, D.; Ross-Ibarra, J. The Potential Role of Genetic Assimilation during Maize Domestication. *PLoS One* **2017**, *12*, e0184202.

Matsuoka, Y.; Vigouroux, Y.; Goodman, M. M.; Sanchez, G. J.; Buckler, E.; Doebley, J. A Single Domestication for Maize Shown by Multilocus Microsatellite Genotyping. *Proc. Natl. Acad. Sci.* **2002**, *99*, 6080–6084.

Mir, C.; Zerjal, T.; Combes, V.; Dumas, F.; Madur, D.; Bedoya, C.; Dreisigacker, S.; Franco, J.; Grudloyma, P.; Hao, P. X.; Hearne, S.; Jampatong, C.; Laloë, D.; Muthamia, Z.; Nguyen, T.; Prasanna, B. M.; Taba, S.; Xie, C. X.; Yunus, M.; Zhang, S.; Warburton, M. L.; Charcosset, A. Out of America: Tracing the Genetic Footprints of the Global Diffusion of Maize. *Theor. Appl. Genet.* **2013**, *126*, 2671–2682.

Perry, L.; Sandweiss, D. H.; Piperno, D. R.; Rademaker, K.; Malpass, M. A.; Umire, A.; De La Vera, P. Early Maize Agriculture and Interzonal Interaction in Southern Peru. *Nature* **2006**, *440*, 76–79.

Piperno, D. R.; Ranere, A. J.; Holst, I.; Iriarte, J.; Dickau, R. Starch Grain and Phytolith Evidence for Early Ninth Millennium B.P. Maize from the Central Balsas River Valley, Mexico. *Proc. Natl. Acad. Sci. U.S.A.* **2009**, *106*, 5019–5024.

Piperno, D. R.; Flannery, K. V. The Earliest Archaeological Maize (*Zea mays* L.) from Highland Mexico: New Accelerator Mass Spectrometry Dates and their Implications. *Proc. Nat. Acad. Sci. U.S.A.* **2001**, *98*, 2101–2103.

Ramos-Madrigal, J.; Smith, B. D.; Moreno-Mayar, J. V.; Gopalakrishnan, S.; Ross-Ibarra, J.; Gilbert, M. T. P.; Wales, N. Genome Sequence of a 5,310-Year-Old Maize Cob Provides Insights into the Early Stages of Maize Domestication. *Curr. Biol.* **2016**, *26*, 3195–3201.

Sigmon, B.; Vollbrecht, E. Evidence of Selection at the Ramosa1 Locus during Maize Domestication. *Mol. Ecol.* **2010**, *19*, 1296–1311.

Stitzer, M. C.; Ross-Ibarra, J. Maize Domestication and Gene Interaction. *New Phytol.* **2018**, *220*, 395–408.

van Heerwaarden, J.; Doebley, J.; Briggs, W. H.; Glaubitz, J. C.; Goodman, M. M.; Sanchez Gonzalez, J. J.; Ross-Ibarra, J. Genetic Signals of Origin, Spread, and Introgression in a Large Sample of Maize Landraces. *Proc. Natl. Acad. Sci. U.S.A.* **2011**, *108*, 1088–1092.

Vigouroux, Y.; Mitchell, S.; Matsuoka, Y.; Hamblin, M.; Kresovich, S.; Smith, J. S. C.; Jaqueth, J.; Smith, O. S.; Doebley, J. An Analysis of Genetic Diversity across the Maize Genome Using Microsatellites. *Genetics* **2005**, *169*, 1617–1630.

Wang, R. L.; Stec, A.; Hey, J.; Lukens, L.; Doebley, J. The Limits of Selection During Maize Domestication. *Nature* **1999**, *398*, 236–239.

Wang, H.; Nussbaum-Wagler, T.; Li, B.; Zhao, Q.; Vigouroux, Y.; Faller, M.; Bomblies, K.; Lukens, L.; Doebley, J. F. The Origin of the Naked Grains of Maize. *Nature* **2005**, *436*, 714–719.

Woodhouse, M. R.; Schnable, J. C.; Pedersen, B. S.; Lyons, E.; Lisch, D. Following Tetraploidy in Maize, a Short Deletion Mechanism Removed Genes Preferentially from One of the Two Homeologs. *PLoS Biol.* **2010**, *8*(6), e1000409. doi:10.1371/journal. pbio.1000409.

Xue, S.; Bradbury, P. J.; Casstevens, T.; Holland, J. B. Genetic Architecture of Domestication-Related Traits in Maize. *Genetics* **2016**, *204*, 99–113.

Yamasaki, M.; Wright, S. I.; McMullen, M. D. Genomic Screening for Artificial Selection during Domestication and Improvement in Maize. *Ann. Bot.* **2007**, *100*, 967–973.

CHAPTER 4

Maize Ideotype

ABSTRACT

Plant breeders search for the plant types with the best plant features, morphology, plant architecture, leaf types, orientation, leaf area, and other characters that promote the crop yield. This chapter briefly discusses various factors affecting maize plant architecture and research activities carried out to improve the ideotype concept in maize.

4.1 ARCHITECTURE OF MAIZE PLANT

The plant architecture is a key determining factor of grain yield and maize has been utilized as a genetic model to study the regulation of plant architecture formation. Current researches unveiled that the phytochrome-mediated red/far-red light signaling pathway and the miR156/SQUAMOSA-PROMOTER BINDING PROTEIN-LIKE (SPL) regulatory module together control the various aspects of plant architecture. Accumulated evidences showed that *ZmSPL* genes perform a significant role in regulating maize flowering time, plant height, tilling, leaf angle, tassel and ear architecture, and grain size and shape. In this study, different ways to utilize maize *SPL* genes and downstream targets for improving maize plant architecture were discussed (Wei et al., 2018b).

In maize, the plant architecture is developed from a balance in the cell fates determining the development of indeterminate or determinate forms. To a large extent, the cells which contribute to indeterminate fates are localized in the meristems, much of these are pluripotent cells. These pluripotent cells are involved in the production of lateral organs. Those meristematic cells which were found at the intercalary stem tissue are responsible in providing the cells for the internodal growth, while those existing along the leaf margins are mostly responsible for bringing about an increase in the leaf width. Zhang et al. (2018a) have identified a mutant in maize that exhibited a defect between the determinacy and indeterminacy balance. This mutant of maize had shorter internodes with more narrowed leaves. They mentioned that the production of these narrowed leaves in this mutant was due to a decline in indeterminate cells in the leaf and also in the stem. In particular, in the shoot meristems, these mutants were a complete failure in controlling the indeterminacy. Though the meristems of the inflorescence were fasciated and determinate in nature, those of axillary meristems were indeterminate. They identified a responsible gene for this type of architecture in maize that was the growth-regulating factor, interacting factor1 (gif1), through the analysis of positional cloning. The mRNA transcripts of this gene were found to be accumulated to a large extent at discrete domains of the meristems in the shoot. In addition, through RNA-seq analysis they found that those genes involved in the architecture of the maize inflorescence were also differentially expressed in the gif1. Further, through the analysis of chromatin immunoprecipitation, they identified that only a few genes have expressed differentially implying as the direct targets of the GIF1.

Maize is a vital food, feed crop, and raw material for the food and energy industry. Optimization of plant architecture plays key role in maize yield improvement. PIN-FORMED (PIN) proteins regulate auxin spatiotemporal asymmetric distribution in several plant developmental processes. Overexpression of ZmPIN1a in maize increased the number of lateral roots (LR) and reduced their elongation, forming a developed root system with longer seminal roots and denser LR. Further, it decreased plant height, internode length, and ear height. This alteration of the maize phenotype improved the yield under high-density cultivation conditions, and the developed root system enhanced plant resistance to drought, lodging, and a low-phosphate environment. Determination of indole-3-acetic acid (IAA) concentration, transport capacity, and application of external IAA showed that overexpression of ZmPIN1a led to increased IAA transport from shoot to root. The buildup of auxin in the root facilitated the plant to apportion more carbohydrates to the roots, improved the growth of the root, and enhanced plant resistance to environmental stress. These results make evidence that root breeding could improve maize plant architecture to create an ideal phenotype for further yield increases (Li et al., 2018).

Thompson et al. (2015) in two divergent populations of maize mapped the genetic architecture of their shoot meristems. They found that the meristem architecture is two divergent populations of maize is population specific. Their study revealed that there was a linkage between the architectures of differentiated and undifferentiated plant organs in maize. Though 15 loci were mapped across these two populations, there was only one identifiable locus in both of these divergent populations of maize.

4.1.1 MAIZE HEIGHT UNDER POLYGENIC CONTROL

Maize plants vary in height. This trait along with other traits is under genetic control. It is an easily identifiable and measurable heritable trait that is predicted much accurately during the estimations of pedigree or the genomic identity of the maize lines. This trait is often variable and is a challenge to the biotechnologists to map the alleles which explain the variations in the maize height. This was addressed by Peiffer et al. (2014). In 7,300 inbreds of maize grown in 64,500 plots across 13 varied environments; they measured the parameters of plant height, the ear height, the time of flowering, and the number of nodes produced. These inbreds represented the maize that was publically available in the United States and also to the families of the maize Nested Association Mapping panel. They performed the joint-linkage mapping of quantitative trait loci (QTL), fine mapping in near-isogenic lines (NILs), genome-wide association studies (GWAS), and genomic best linear unbiased prediction (GBLUP). Through these studies, it was found that the heritability of maize height was around 90%. There was a large variation in plant height in the maize lines of NAM family-nested QTL. The variation in height ranged from 2.1–0.9%. Validation of two tropical alleles by fine mapping at this QTL in NIL families indicated that there was a considerable linkage by GWAS colocalized with established height loci, comprising brass-inosteroid-deficient dwarf1, dwarf plant1, and semidwarf2. Further, GBLUP explained 80% of height variation in the panels. It also outperformed bootstrap aggregation of family-nested QTL models in evaluations of prediction accuracy. Thus, these research findings have indicated that maize height architecture is under high polygenic control and is heritably under strong genetic control.

4.1.2 MAIZE LEAF ARCHITECTURE

Flowering time in maize is one of the factors enabling maize to adapt to the local environmental conditions. This flowering time is measured by the number of leaves that are produced on the plant. On the whole, the plant architecture of maize is determined by the number of leaves and also their distribution pattern on the plant. Li et al. (2016) genotyped BC 2S3 recombinant 866 lines of maize-teosinte, through the usage of 19,838 single nucleotide polymorphism markers to dissect and evaluate the genetic architecture of leaf number in maize and to gain a comprehensive understanding of its genetic relationship with the flowering time. Their findings have shown that of the total leaf number, in maize, those leaves which are above the primary ear and those below the primary ear are considerably under independent genetic control. They suggested that during domestication and improvement, these might have been subjected to differential directional selection. They revealed that two traits, namely, the flowering time and the leaf number in maize are usually regulated at a modest level. Furthermore, they have revealed that in these inbred lines, flowering time and leaf number are usually regulated at a modest level. In these, by near isogenic analysis, they have validated the pleiotropy of the genes ZCN8, dlf1, and ZmCCT on the leaf number and also on flowering time. Fine mapping of these genes has revealed that qLA1-1, was a key-effect locus found to be that distinctively affecting the leaf number above the ear. This locus was restricted to a region with strict recombination suppression that was obtained from teosinte. Thus, they highlighted that in maize the genetic independence of leaf above ear from leaf below ear enable in optimizing the leaf number leading to the production of ideal plant architecture suitable for maize breeding.

Leaf architecture has a direct effect on canopy structure, consequentially influencing yield. A maize mutant, drooping leaf1 (drl1) with unusual leaf architecture was discovered. It was found that the pleiotropic mutations in drl1 influence leaf length and width, leaf angle, and internode length and diameter. Another allele drl2 produced by transposon mutagenesis was located at the drl2 enhancer locus. This drl2 allele was found to improve the phenotype and interacted synergistically with drl1 mutants. The drl genes were co-expressed in emerging and growing leaf primordia at the shoot apex, but not in the vegetative meristem or stem. GWAS with maize NAM-RIL populations showed that the drl loci exist within quantitative trait locus regions for leaf angle, leaf width, and internode length. Thus, this study reveals that drl genes control the development of significant agronomic characters in maize (Strable et al., 2017).

The gene expression is regulated by a small part of the genome sequence, but the biochemical identification of this fragment is costly and strenuous. In species like maize, this is more difficult because the varied intergenic regions and a large number of repetitive elements restrict the usage of the data from one line to the other. Although regulatory regions are few, they do have distinguishing chromatin backgrounds and sequence organization (the grammar) with which they can be recognized. In this study, a computational outline was developed to make use of this sequence arrangement. They constructed "bag-of-k-mers" and "vector-k-mers" models that can differentiate the regulatory and nonregulatory regions with 90% accuracy. The "bag-of-k-mers" attained higher overall precision, whereas the "vector-k-mers" models were more helpful in detecting the important sequence groups within the regulatory regions. Therefore these models offer potent tools to interpret regulatory regions in other maize lines beyond the reference, at low cost and with high precision (Mejía-Guerra and Buckler, 2019).

In plants, the nitrogen (N) uptake is regulated by root architecture. The current methods of root architecture phenotyping are restricted to seedlings or the outer roots of mature root crowns. The functional integration of root phenes is not studied clearly. Phenes and phene modules associated with N uptake were discovered in maize by conducting intensive phenotyping of mature root crowns. Under full and deficient N regimes 12 maize genotypes were grown. Using custom software root phenes comprising nodal occupancy, angle, diameter, distance to branching, lateral branching, and lateral length were measured. The phenes associated with size such as diameter and number were found to be considerably influenced by nodal position. It was suggested that the greater distance to branching, namely, the length from the shoot to the emergence of laterals, is a novel phene that reduces placing roots in previously explored soil. It was found that the root phenes from both old and young whorls of nodal roots contribute to variation in shoot mass and N acquisition. Thus, these studies demonstrated the usefulness of intensive phenotyping of mature root systems and the significance of phene integration in soil resource uptake (York and Lynch, 2015).

To increase the yield and total nitrogen-use efficiency (NUE) of maize, a greater understanding of nitrogen uptake and its use is essential. Here, two distinct maize inbred lines, B73 and F44 were examined for their variances in N uptake and utilization abilities. They used transcriptional, enzymatic, and nitrogen transport analytical tools for the study. The results revealed that the genetic, enzymatic, and biochemical pathways of root nitrogen transport and assimilation vary significantly in B73 and F44. Further, B73 exhibited a better capacity for ammonium absorption and transport, while F44 favored nitrate. These differences in inbred lines were found to be associated with RSA. F44 was characterized with longer crown roots (CR), higher surface area, and volume with more LR than B73. The B73 had abundant primary, seminal, and CR but lacked the characteristics of the F44 CR. Thus, through their research findings, they proposed that nitrogen management is associated with the RSA and both variables should be considered while breeding maize varieties for high NUE (Dechorgnat et al., 2018).

Pautler et al. (2015) identified fasciated ear 4 (fea 4) a semidwarf mutant in maize producing fasciated ears and tassels as a bZIP transcription factor orthologous to PERIANTHIA of *Arabidopsis thaliana*. They found that this transcript FEA4 expressed in the peripheral zone of the vegetative shoot apical meristem and also in the vasculature of immature leaves. Its expression was excluded in the stem cell niche at the tip of the shoot apical meristem and from the leaf primordia which were in the incipient stage of development. The expression of FEA4 was detected all through the entire inflorescence and floral meristems when the shoot apical meristem transitioned to the reproductive fate. This pattern of expression was recapitulated with the native expression of a functional YFP:FEA4 fusion. They identified 4,060 genes proximal to the FEA4 binding sites. Their findings have suggested that FEA4 promotes the differentiation in the meristem periphery through its influence on the regulation of the responses which were auxin-based along with those genes connected with the differentiation and polarity of leaf.

The wild ancestor of maize, teosinte is highly branched. In the process of domestication of maize, the suppression of the axillary branch has been attained through the selection of a gain of function allele of the teosinte branched1 (tb1) transcription factor. This factor acted as a repressor of axillary bud growth. Earlier research findings have shown that other loci may also function epistatically with tb1 and might also be accountable for some of its phenotypic effects. Dong et al. (2017) have shown that this tb1 mediated the suppression of the axillary branch, through its direct activation of the tassels to replace the upper ears 1 (tru1) gene. This gene encoded an ankyrin repeat domain protein, which contained a BTB/POZ motif. This motif was indispensable for protein–protein interactions. The expression of TRU1 and TB1 was overlapped in axillary buds. Analysis by chromatin immunoprecipitation and gel shifts has revealed that TB1 binds to two locations in the tru1 gene. Additionally, the nucleotide diversity surveys have indicated that tru1, like tb1, was also a target of selection. TRU1 has its expression largely in leaf vasculature of axillary internodes in modern maize; however, the expression of this in teosinte, was absent or largely declined. Thus, they mentioned that a chief innovation that led to the creation of an ideal plant architecture in maize had its origin from the ectopic overexpression of tru1 in axillary branches, a significant step responsible for mediation of the effects of domestication by tb1.

4.1.3 SEMIDWARF GENES IN MAIZE

Identification of semidwarf genes and their incorporation into the crop plants of rice and wheat in the 1960s resulted in an effective reduction in plant stature, lodging, and improved the yields during this green revolution. Wei et al. (2018a) identified a novel allele Brachytic 2 gene (qpa1) involved in encoding of P-glycoprotein, affecting the plant stature of maize without negative effect on yield. This allele was identified through a modified method of fine mapping. This gene had a 241-bp deletion in the last exon. The expression of gene not only resulted in modified plant architecture, there was also a decline in the plant and ear height, with an increase in the stalk diameter and production of more erect statured leaves. This gene also had an influence on the differential co-expression of four genes affecting plant height, namely, D3, BAK1, Actin7, and Csld1. These genes are functional in the biosynthesis of gibberellin and brassinosteroid, auxin transport, and cellulose synthesis. Thus, the research findings have indicated that qpa1 can be utilized for competently modifying the plant stature in maize along with the combined application of D3, BAK1, Actin7, Csld1, and the other 95 differentially co-expressed genes, which can be edited by means of novel genomic editing tools.

4.1.4 GENETIC ARCHITECTURE

Linkage and genome-wide association analyses of the genetic architecture of a population of recombinant inbred lines for flowering and leaf traits from the inflorescence of the male and female inbred lines by Brown et al. (2011) has revealed that the loci of inflorescence had a larger effect rather than the loci of the leaf or flowering. Similarly, than tassel effects, the effects of the ear appeared large. The ability of ear trait models in predictability was much lower than the predictive abilities of the models of tassel, flowering or those of leaf trait models. The studies revealed many pleiotropic loci controlling the ear and tassel elongation, with a similar developmental origin, even though there were identical polymorphisms, the pleiotropic ear effects are to an extent larger than those of the tassel effects, revealing that these variations in the genetic architecture are mainly a function of trait stability over the evolutionary time.

An analysis of genetic architecture of maize traits in 336 lines of maize through phenotyping and genotyping with single nucleotide polymorphisms of Bouchet et al. (2017) revealed 34 QTLs for individual traits and for trait combinations 6 QTLs. Out of these QTLs, only five were pleiotropic in nature. They detected a cluster of these QTLs to be located in a 5-Mb region around Tb1 which is generally related to the tiller number, first PCA axis and to ear row number. Further, it was found that there was a positive association between the first PCA axis and flowering time with a negative correlation to that of yield. They identified that for tillering the candidate genes were Kn1 and ZmNIP1, while for the Rubisco Activase 1 and for leaf number the candidate gene was ZCN8. Variations in the QTLs were larger for those traits which are closely related to the plant development, namely, tillering, flowering time. These variations in the QTLs were less related to those traits of growth, namely, the yield components.

4.1.4.1   INSIGHTS INTO MAIZE TASSEL GENETIC ARCHITECTURE

During the process of domestication and improvement maize has undergone intense changes in the morphological characters of the plant. Many researchers have identified several genes that have a role in the development of the maize inflorescence. To understand and to have an insight into the genetic base of morphological changes that have occurred in maize tassel development during the process of domestication, Xu et al. (2017) evaluated a huge population (866) of maize-teosinte BC2S3 recombinant inbred lines. These inbred lines were further genotyped by the use of single nucleotide polymorphism markers (19,838). Further, five tassel morphological traits were mapped under high resolution quantitative trait locus mapping. This mapping has shown that these traits are connected closely with many varied genetic architectural features involved in the inflorescence development. They also identified some of the known genes through mutagenesis, which were involved in the development of inflorescence. These known genes were considerably enriched in the QTLs of the tassel trait. Most of these genes, which included ramosa1 (ra1), barren inflorescence2 (bif2), unbranched2 (ub2), zeafloricaula leafy2 (zfl2), and barren stalk fastigiate1 (baf1), have shown the evidence of selection. An in-depth nucleotide diversity analysis that was carried out at the bif2 locus led to the identification of a well-built selection signatures in the 5'-regulatory region. Additionally, the research findings have revealed that most of the genes involved in flowering time are colocalized with tassel trait QTLs. An additional supplementary analysis of association analysis has shown that ZmCCT the photoperiod gene of maize was considerably linked with the variations in the tassel size. Using near isogenic lines, they have narrowed down a most significant effect QTL for tassel length, qTL9-1, to a 513-kb physical region. Their findings have provided key visions into the genetic architecture involved in controlling the maize tassel morphological changes during the development and evolution process.

GWAS of maize nested association mapping panel, at US has revealed that the key contributor of eight fold increase in maize yields in the past 70–80 years was due to alterations in the plant architecture, particularly in the leaf angle and leaf size, which allowed an effective light capture for assimilate synthesis. Much of the gain in yields was attained on the basis of selection of the above parameters in breeding new lines by the breeders. In addition, an increase in the planting densities has also resulted in the production of higher yields. Feng et al. (2011) analyzed the genetic basis of key leaf traits governing the leaf architecture and leaf angle. Their results have revealed that most of these genetic architecture governing leaf traits had less epistasis or the pleiotropic and environmental interactions. In particular, changes in the liguleless genes in the population resulted in the production of more upright leaves.

Many studies have shown that the yields are to a large extent influence by canopy structure, which in turn is under the direct influence of leaf architecture. Strable et al. (2017) identified a mutant, of maize (drooping leaf 1, drl1) possessing aberrant leaf architecture. In this mutant, the pleiotropic mutations that occurred had their effect on the leaf length, width, angle as well as internodal length and diameter in plants. In the Mo17 inbred line, they found that a natural variation has occurred at the drl2 enhancer locus that resulted in the declined expression of drl2-Mo17 alleles, and the phenotypes that resulted were highly enhanced. Another allele (drl2) has been produced with the transposon mutagenesis. This allele exhibited synergistic interactions with the drl1 mutants and has resulted in a decline in the expression of transcript levels of drl2. These findings have indicated that in maize, for the attainment of a proper and desired leafing pattern and the development and the proliferation of leaf supporting tissues, as well as for the restriction of the expansion of the auricle at the region of leaf midrib, the requisite of the presence of drl genes is essential. In maize, the CRABS CLAW coorthologs are encoded largely by the paralogous loci, belonging to the YABBY family of genes acting as transcriptional regulators. Coexpression of drl1 genes was seen to a large extent in the budding and emergent leaf primordia at the region of shoot apex. Their expression was not detected in these incipient leaf primordia arising at the vegetative meristem or meristems of stem. In NAM- RIL populations, the genome-wide studies have shown that the loci of drl reside within the regions of quantitative trait locus for leaf angle, leaf width, and internode length. For traits of leaf width and internodal length, in maize, the leaf has a proximal sheath, enabling in stem grip and a distal long leaf blade. Both the leaf sheath and the leaf blade are attached by an auricle and a membranous ligule. This hinge-like region enables the leaf blade and the supporting midrib to bend far away from the stem. This bending gives maize an architecture that has an effect on the yield (Lockhart, 2017). In general, widespread analyses of *liguleless* (*lg*) mutants have revealed the important roles of the *lg* genes in establishing the blade-sheath boundary (Johnston et al., 2014), though their downstream targets were yet to be revealed. Strable et al. (2017) discovered another highly informative maize mutant, *drl1*. The leaves of *drl1* droop downward due to decreased midrib development and show increased leaf angle because of distal extension of the auricle along the blade-sheath boundary. In addition, *drl1-R* (the reference allele of *drl1*) has shorter, narrower leaf blades and longer proximal sheaths than the wild type, along with long, slender internodes. These phenotypes are specifically marked in the existence of the unlinked genetic modifier *drl2-Mo17*. Similarly, the Transposon mutagenesis has discovered a second allele of *drl2* that interacted synergistically with *drl1* in a dose-sensitive manner. The maize *drl1* and *drl2* loci encode CRABS

CLAW-like transcription factors in the YABBY family (Bowman and Smyth, 1999). The *YABBY* genes were found to have vital roles in the development of leaf lamina in many species, though their role in maize has yet to be clarified. However, Juarez et al. (2004) has suggested that these genes are helpful in directing the outgrowth of the leaf blade. Lockhart (2017) by means of a nested association mapping population (NAM) of ~5000 recombinant inbred lines, by GWAS determined whether the two loci, namely, *drl1* and *drl2* loci colocalize with some of the agronomically significant QTL. Further, an analysis of the association of SNPs (single nucleotide polymorphisms) with the earlier generated phenotypic data that arose from the natural variation in the NAM population has revealed that *drl1* and *drl2* are indeed located in important QTL regions for stem and leaf traits and that quite a lot of unusual SNPs have large phenotypic effects on these traits have recognized unusual single nucleotide polymorphisms, having huge phenotypic effects.

Similarly, in another study, Lewis et al. (2014) revealed that LG1 accumulates at the site of ligule formation and in the axil of developing tassel branches. The dominant mutant Wavy auricle in blade1 (Wab1-R) produced ectopic auricle tissue in the blade. It also increased the domain of LG1 accumulation. Positional cloning and revertant analysis have revealed that wab1 encoded a TCP transcription factor. Tassel branches were less and more in upright stature in the wab1 revertant tassel. In the dominant mutant, these tassel branches had an increased branch angle. They detected that wab1 mRNA was expressed at the base of branches in the inflorescence. This expression was required for the expression of LG1. Though the expression of wab1 was lacking in the leaves, its expression was found to be present in the dominant mutant. They mentioned that though wab1 is not obligatory for induction of lg1 expression in the maize leaf, its expression is required for counteracting the severe phenotype of the dominant Wab1-R mutant. Thus, this regulatory interaction of LG1 and WAB1 revealed the presence of a link between that of maize leaf shape and the architecture of tassel and has suggested that ligule acts as a boundary identical to that present at the base of the lateral organs.

In maize, perfect plant architecture, with an erect leaf angle, optimum leaf orientation is more suited for the effective light capture by the canopy and proper air circulation between the leaves, particularly when maize is planted under high-density plantings. Li et al. (2015), QTL mapping and joint linkage analysis in 538 RILs of maize for leaf architecture, have revealed the presence of 45 QTLs with a large range of phenotypic effects (1.2–29.2%). These were identified from among these inbred lines of six environments for leaf angle traits. They also identified, by joint linkage mapping, four leaf architecture traits. Almost all these QTLs of each trait were able to explain about 60% of the phenotypic variance. Four QTLs were found on small genomic regions where candidate genes were detected. Genomic estimates from a GBLUP model elucidated 45 ± 9% to 68 ± 8% of the variation in the remaining RILs for the four traits.

The phenotypic key trait contributing to productivity levels of maize is the presence of an erect leaf angle in the architecture of the plant. The presence of an erect leaf angle not only enables in bringing about an increase in the light harvesting of incident radiation for effective photosynthesis at the vegetative stage but also during the grain filling stage. Vast information is available on the QTL controlling leaf angle of maize. Ku et al. (2010) detected the genetic control of maize plant architecture of narrower leaf angle and optimum leaf orientation value in a set of 229 $F_{2:3}$ families which have been obtained from the cross between compact and expanded inbred lines. They mapped the QTLs and evaluated these lines at three varied environments. Out of the 20 QTLs that were detected, only 5 QTLs could explain about 53.9% of the phenotypic variance for narrowed leaf angle and optimum leaf orientation. They identified two major genome regions involved in the control of leaf angle and leaf orientation, qLA1 and qLOV1 at adjacent marker umc2226 on chromosome 1.02 accounted for 20.4 and 23.2% of the phenotypic variance, respectively; qLA5 and qLOV5 at nearest bnlg1287 on chromosome 5 accounted for 9.7% and 9.8% of the phenotypic variance, respectively. These findings were valuable for marker-assisted selection in enhancing the performance of maize plants and improved yields under high-density plantings for increasing the amount of light capture per unit area of land.

Wei et al. (2016) in a study identified the QTL for traits associated with maize leaf area. They found these QTLs across three environments in a set of recombinant inbred lines, through the usage of 1226 single nucleotide polymorphic markers. Though they have detected 16 QTLs, only four of the QTLs had a larger effect of more than 10%. Among these, five of the QTL elucidated 46.02%, seven of them elucidated 46.77%, and four QTL elucidated 30.03% of the phenotypic variance of maize leaf length, width and area. They identified additional epistatic effects for all of the maize chromosomes, except for chromosomes 7, 8, and 9. Almost all of these epistatic effects had an involvement of pairs of loci on dissimilar chromosomes.

Similarly, the molecular mechanism of the QTLs was elucidated in a study by Zhang et al. (2014). Through fine mapping and positional cloning, they identified the gene ZmCLA4, of the qLA4-1 QTL linked with the leaf angle of maize. The study revealed that it was a close ortholog of the gene LAZY1 of rice and Arabidopsis. The sequence analysis of this gene has discovered the presence of two SNPs and two indel sites in this gene between D132 and D132-NIL inbred lines of maize. Further, there was a strong association with leaf angle by the C/T/mutation 667 and CA/ indel 965 as revealed by association analysis. The nuclear localization signal in ZmCLA4 was confirmed to be located in the qLA 4 QTL which regulated the leaf angle. The transcript accumulation of this gene was more in D132-NIL than D132 inbred line. Their results also revealed that this gene exhibits a negative role in regulating the leaf angle of maize, by altering the accumulation of the mRNA, resulting in the distorted shoot gravitropism and cell development.

At high plant densities, maize plants possessing an upright architecture have increased the maize yields dramatically. Tian et al. (2019) suggested that introgression of the wild UPA2 allele into modern hybrids and editing ZmRAVL1 would improve these yields. They have cloned UPA1 (Upright Plant Architecture1) and UPA2, two QTLs that were conferring upright plant architecture. They detected that UPA2 is under the control of a two-base sequence polymorphism that regulated the expression of a B3-domain transcription factor (ZmRAVL1). This transcription factor was located 9.5 kilobases downstream. The UPA2 exhibited differential binding by DRL1 (DROOPING LEAF1). The DRL1 physically inter-acted with LG1 (LIGULELESS1) and repressed the LG1 activation of ZmRAVL1. ZmRAVL1 regulated the brd1 (brassinosteroid C-6 oxidase1), which underlies UPA1, altering endogenous brassinosteroid content and leaf angle of maize. Thus, the UPA2 allele that was found to decline the leaf angle of maize had its origin from the wild ancestor of maize, the teosinte and was lost during the domestication of maize.

In maize, plant architecture is influenced by leaves and in turn, these have an influence on crop yields. Wide-spread transcriptional profiling of the maize leaf tissues, namely, leaf blade, ligule, and sheath has been conducted by many researchers. Dong et al. (2019) dissected the dynamics of transcriptome of leaf sheath tissues supporting the leaf blade. A wide range of data sets were used for the study to unveil the importance of these in supporting leaf blade. At the stage of maturity, a dynamic change in the leaf sheath transcriptome was evident. This transcriptome change was due to the buildup of basic cell structures at the initial and early stages of leaf sheath maturation and these later transitioned to synthesis in the cell-wall components and other modifications. This transcriptome has revealed a change at the last stage of the maturation of the leaf sheath in relation to photosynthesis and lignin biosynthesis. In maize, the biological functions of various tissues are largely specialized. A higher expression of 15 genes in leaf sheath was identified. However, the expression of these genes was much lower in that of the leaf blade. These included the BOP2 homologs GRMZM2G026556 and GRMZM2G022606, DOGT1 (GRMZM2G403740) and transcription factors from the B3 domain, C2H2 zinc finger, and homeobox gene families, associating these genes in sheath maturation and organ specialization.

4.1.5   ROOT ARCHITECTURE

In maize, root initiation takes place immediately after seed germination and as the seedling development takes place, there will be the development of embryonic and nodal root systems. The early embryonic root system in maize comprises a primary root (PR) and many seminal roots, while the later formed nodal root system has many aerial nodal roots that have arisen from the above nodal regions of the shoot. Most of the CR are formed from the belowground nodes. Thus, the uptake of water and mineral nutrients at the seedling stage is through the early embryonic root system, while that at the later stages is through the nodal root system. The presence of aerial roots on the lower nodes also enables in providing support to the plant and keeps the plant in upright positions, as well as influence the lodging resistance.

The plant's ability in accessing soil water is related to the architecture of root systems. The root architecture also exhibits its influence in increasing the plant's adaptation capacity to the conditions of water limitations. In general, the phenotypic screening studies take into consideration the root attributes of young plants as the main criteria of screening trait. Singh et al. (2010) characterized the root architectural development (morphological), its development, and time of origin of roots in hybrids of maize. The study revealed that maize produces three to seven seminal roots comprising of the primary and scutellum roots and that of the coleoptile nodal roots when it is at the second leaf stage. The absorptive capacity of a plant is dependent on the complex organization of the root system. Maize root architecture, cell-cycle proteins, and production of heat-shock proteins were studied with reference to the auxin-induced changes in seedlings of maize which were raised from the embryonic axes by Martínez-de la Cruz et al. (2015). Their results have shown that in the initial stages, namely, during the first few days after germination, the seedlings of maize have developed many types of root types exhibiting a simultaneous growth. However, it was observed that its post-embryonic root development had begun with the development of the PR and seminal scutellar roots. Later there was the formation of adventitious CR, brace roots (BR), and LR. The root architecture of maize was influence by auxins in a dose-response pattern. Among the various auxins evaluated, the crown root formation was stimulated to a large extent by NAA and IBA. A large repressive effect on root growth was seen with the application of 2,4-D. The levels of cell cycle proteins CKS1, CYCA1, and CDKA in root and HSP101 were modulated in their expression and regulated the timing of early branching patterns of maize root development under the influence of auxins.

In the maize root system, the root types vary in structure and function. Hochholdinger et al. (2018a) through mutant analyses have identified that there are a few root-specific genetic regulators that intrinsically control the architecture of the root system of maize. Molecular cloning of these genes has shown that some of the chief elements involved in auxin signal trans-duction, namely, LBD domain and Aux/IAA proteins, are key instrumental elements for the initiation of seminal, shoot-borne, and lateral root initia-tion. In addition, the genetic analyses have confirmed that genes connected to exocytotic vesicle docking, cell wall loosening, and cellulose synthesis and organization also regulate the elongation of root hair. The detection of upstream regulators, protein interaction partners, and downstream targets of these genes together with cell-type-specific transcriptome analyses have made available new insights into the regulatory networks involved in the control of root development and root architecture of maize.

Salvi et al. (2016) in a group of introgression lines of maize obtained from a reference line B73 and the landrace Gaspe Flint dissected the genetic architecture and root diversity in these lines during the early stages of root development. They have identified three QTLs involved in the control of seminal root numbers on the chromosome bins 1.02, 3.07, and 8.04–8.05. They could explain 66% of the phenotypic variation in these lines. Further, it was revealed that the allele for the lowest seminal root number in all the three chromosome bins was largely contributed from Gaspe Flint. A negative correlation was found to exist between the dry weight of PR and the seminal root number ($r = -0.52$). Furthermore, co-mapping of QTLs for PR size with seminal root number QTLs, has shown that these seminal root number QTLs had a pleiotropic effect on the seminal root number of PR. They mentioned that this pleiotropic effect might have been resulted due to the competition for seed resources. Interestingly, two out of three SRN QTLs co-mapped with the only two known genes (rtcs and rum1) of maize, that were affecting the number of seminal roots have indicated the presence of a strong additive effect of the three QTLs. They mentioned that the development of NILs for each QTL in the elite B73 background enables the provision of distinctive opportunities for characterizing the functional genes involved in the root development of maize. It also provides an opportunity to assess the effects of root architectural traits on the seedling establishment, early development, and maize yields.

Giuliani et al. (2005) have found that in near-isogenic hybrids of maize, differing at a key QTL for leaf ABA concentration, under different water regimes, that this QTL has its influence on affecting the root lodging through a constitutive effect on maize root architecture. Pace et al. (2015) in a genome-wide association analysis of maize analyzed the allelic diversity of complex traits and identified superior alleles. They have genotyped 384 inbred lines from the Ames panel with 681,257 single nucleotide polymorphism markers using genotyping-by-sequencing technology. Further, they have phenotyped 22 seedling root architecture traits.

Hochholdinger and Tuberosa (2009) identified many genes by mutant analysis in maize, involved in the regulation of shoot-borne root initiation and also the root hair elongation (RTH1 and RTH3). Further, through QTL studies they have emphasized the role of seminal, LR, and root hairs in the maize root architecture in the acquisition of phosphorus from the soil. Moreover, they have shown that the QTLs which were influencing the root features also had their influence on the yield attributes at varied conditions of water regimes. Additionally, they have also provided insights into the root development. With the help of the analyses of proteome and transcriptome, they have also identified few key candidate genes that were connected to the specification of root cells and initiation of LR in the cells of pericycle.

Phenotypic profiling of 187 advanced-backcross BC4F3 maize lines of (Ye478 × Wu312) at different developmental stages under field conditions by Cai et al. (2012) has shown that there was a higher rate of heritability for root traits of axial root-related ones than the traits of the lateral root. It has indicated that lateral root growth is more influenced by environmental effects. Stage I, namely, root establishment had a closer relationship with grain yield. They detected 30 QTLs for RSA in this BC4F3 population. In the maturation stage, only 13.3% of QTLs were detected. It was found that most of these QTLs were located on chromosome 6 near the locus umc1257 (bin 6.02–6.04) at the establishment stage, on chromosome 10 near the locus umc2003 (bin 10.04) for the number of Axial roots across all three developmental stages. The regions of chromosome 7 near the locus bnlg339 (bin 7.03) and chromosome 1 near the locus bnlg1556 (bin 1.07) harbored QTLs for both grain yield and lateral root related traits at the establishment stage and silking stages, respectively.

Zhang et al. (2018b) analyzed the genetic architecture of maize nodal root system in a teosinte maize population. Their high-resolution mapping has detected a total of 133 QTLs. Out of which 62 QTLs have accounted for more than half of the genetic variation in this population for nodal root number and these derived QTLS were those QTLs related to flowering time. These QTLs were validated also through the analysis of transgenic analysis and a GWAS. The results have shown that only 16% of the total genetic variation found for the nodal root number has been found to be derived from those QTLs for plant height. Thus, their findings have suggested that in maize, the flowering time has a key vital part to play in determining nodal root number via its indirect selection during the domestication of maize. Furthermore, the results have also indicated and supported that in temperate maize, root architecture has the presence of more aerial nodal roots and fewer cown roots, which favors the improvement of temperate maize root lodging resistance and also enhances its ability to extract water and nitrogen present in the deeper layers of soil, particularly under high-density planting systems.

4.1.6 PROTEOMICS IN IDENTIFICATION OF ROOT ARCHITECTURAL CHANGES

Recently, proteomics is very intensively studied and applied for the identification of proteins that are involved in forming the three-dimensional root architecture, with varied structure and functionally diverse roots. Hochholdinger et al. (2018b) studied the proteomic changes that occurred at the initiation or emergence of lateral and seminal roots from the shoots by usage of developmental mutants. Proteomic changes in single-cell-type root hairs were also surveyed. Additionally, these proteomic changes were also surveyed in the PRs for their developmental changes, in dissimilar tissues as the meristematic zone, the elongation zone as well as stele and cortex of the differentiation zones to provide an insight into the multifaceted proteomic interactions of the maize RSA.

4.1.7   VISUAL SCORING OF PHENOTYPES FOR ROOT ARCHITECTURE TRAITS

Trachsel et al. (2011) presented a method of visual score 10 root traits to score the root crown architecture of maize plants in three recombinant inbred line populations, namely, B73xMo17, Oh43xW64a, Ny821xH99 in Pennsylvania. The visual score was based on the score of root numbers, angles and branching pattern of crown and BR. They evaluated these traits in these inbred populations at the flowering stage. They mentioned that time taken for the visual score of a sample was only two minutes. Visual scoring of root crowns have given a reliable estimate of values for these traits of root architecture, specified by high correlations between measured and visually scored trait values for numbers (r^2 = 0.46–0.97), angles (r^2 = 0.66–0.76), and branching (r^2 = 0.54–0.88) of brace and CR. On the basis of this visual evaluation of root traits of CR, they were able to discriminate the populations. The results revealed that more number of roots, with the highest branching density, shallow root angles were more prominent in the RILs which were derived from the cross NY821 × H99. However, in inbred lines of the cross between OH43 × W64a the root angles were very steep. Similarly, better correlations were seen in the visual scoring of the BR rather than that of the CR. Thus, they mentioned that this would be one of the valuable tools assisting in the evaluation of root architecture across specific environments.

4.1.8   MAIZE POPULATION DENSITY AND LIGHT ATTENUATION

The size, shape, and orientation of shoot components determining the canopy architecture has an influence on the amount of light that attenuates within the canopy. Light attenuation in row crop grown plants as maize is to a large extent governed by the canopy architecture. The cultural operations have an influence in improving the efficiency of light interception as they affect and modify the canopy architecture. Maddonni et al. (2001) evaluated in Argentina, the response of four maize hybrids to variable population densities and row spacing in the production of leaf growth as well as light attenuation. Variation in plant population densities in these maize hybrids resulted in large variations in the production of shoot organs. From the initial stages of crop growth, only the influence of population and row spacings were evident on the leaf growth (V_6–V_8) and that of azimuthal orientation (V_{10}–V_{11}). Modification of shoot size and orientation of leaves revealed the prevalence of shade avoidance reactions, which were thought to have been probably due to the triggering by a reduction in the red:far-red ratio of light within the canopy. One of the hybrids displayed a random azimuthal distribution and was designated as a rigid hybrid. However, even at very low plant population densities, there was a modified azimuthal distribution of leaves. This was evident in these hybrids in response to plant rectangularity. Such hybrids were designated as plastic in nature. There was no variation in the light attentuation in these hybrids after the attainment of the maximum leaf area index. However, it was found that a more uniform plant distribution resulted in an increase in light attenuation (k coefficient: 0.37–0.49) only when crop canopies have not attained the critical leaf area index values.

Perez et al. (2019) mentioned that breeders often select for traits contributing to yield. They tested the effect of breeding on a variety of traits which were included in determining the maize plant architecture and light interception, by an analysis in a panel of sixty maize hybrids released from 1950 to 2015. Their study was on the basis of new traits, which were calculated from reconstructions obtained from a phenotyping platform. They assessed the contribution of these traits to light interception in virtual field canopies composed of 3D plant reconstructions, with a model tested in a real field. They found that two categories of traits had varied contributions to genetic progress, namely, (i) the vertical distribution of maize leaf area which exhibited a high heritability and has shown a noticeable trend over generations of selection. The leaf area of maize tended to be positioned at lower positions in the canopy. This was, therefore, responsible to bring about an improvement in the light penetration as well as light distribution within the canopy. Thus, it potentially enlarged the carbon availability to ears, via the amount of light absorbed by the intermediate canopy layer;

(ii) neither the horizontal distribution of leaves in the relation to plant rows nor the response of light interception to plant density has shown any appreciable trends with generations. Thus, they suggested that in maize, among the various traits affecting its plant architecture, the main trait of indirect target of selection is the vertical distribution of leaf area.

4.1.9   ROOT AND CANOPY ARCHITECTURE MODEL

Hammer et al. (2009) through a modeling approach generalized a maize crop model to explain the trends of changes in the architecture of root and canopy architecture on the yield trends of maize in the US Corn Belt. They introduced a layered, diurnal canopy photosynthesis model for predicting the consequences of variations in the architecture of maize canopy and a two-dimensional root exploration model for predicting the changes and consequences of the variations in the architecture of root system and evaluated them in field experiments in base hybrid of maize, Pioneer 3394. Analysis of simulation studies and field studies of these changes in canopy architecture and RSA in a range of sites, soils and densities has shown that any slight variation that occurred in the architecture of the root system in the capturing of water from the soil had a direct effect on the accumulation of maize biomass and yield. However, the change in the canopy architecture exhibited a minor direct effect on the biomass accumulation and yield trends, though its effects are indirect. These indirect effects are seen in the retention of leaf area and the assimilate partitioning to the development of the maize ear.

4.1.10   MODEL FOR LATERAL ROOT BRANCHING DENSITY

Postma et al. (2014) used the SimRoot plant model to simulate a three-dimensional development of maize root architecture with lateral root branching densities and quantified the uptake rates of nitrate and phosphorus and plant carbon balances in soils which exhibited variations in the availability of these nutrients. The observed phenotypic variation in the lateral root branching density of maize is large, that is, 1–41 cm^{-1} of the major axis at specific environments. The results revealed that nitrogen acquisition was favored largely by the sparsely spaced LR (less than 7 branches cm^{-1}), long laterals were optimal for nitrate acquisition. Similarly, densely spaced (more than 9 branches cm^{-1}), short laterals were optimal for phosphorus acquisition. As there exists, a strong competition for nitrate between the LR, it would result in an increase in the branching density of LR and to a decline in the uptake of nitrate per unit root length. The carbon budgets of the plant did not permit larger root length (i.e., individual roots in the high-lateral root branching density plants were shorter). The competition coupled with the limitation of the carbon toward the growth had less influence on the uptake of phosphorus, though there was an increase in the lateral root branching density and larger root length and nutrient uptake. Thus, the results have concluded that in maize, the optimal rate of lateral root branching density is more dependent on the virtual availability of nitrate (a mobile soil resource) and phosphorus (an immobile soil resource) and is larger in those environments which have higher rates of carbon fixation.

4.2 MAIZE IDEOTYPE

Maize is cultivated all globally and is a strategic crop; it has high tolerance to radiation intensities and shows high water use efficiency. So, there is a need to develop maize ideotypes for site-specific conditions with the descriptions of past, present, and future ecological history, and the response of the local material by means of empirical or mechanistic modeling. Genotyping studies revealed that a single trait can not cope with the present climatic variability and can never improve the performance of plants in all climatic scenarios. The ideotype includes several types of biological characters or the genetic basis that confers improved performance for a specific biophysical environment, particular cropping system and end usage of the crop. Earlier, scientists used to consider visual and growth phenotypes for ideotyping, but future ideotyping trend will concentrate more on the genotypic traits. The forthcoming ideotype strategies will be based on biotechnological tools eased by the bioinformatics, filling gaps in the present knowledge and overwhelming the challenges of climatic change and increased the world population (Khakwani et al., 2018).

In recent years, attaining grain supply security with limited arable land and under a continuously changing climate has become a major challenge. Being a C4 plant, maize has a high yield potential and is expected to become the leading cereal crop in the world by 2020. With the exploitation of hybrid and production technologies, maize production reached a plateau in many countries. Therefore, there is a need to develop maize ideotype with traits and architectures that can tolerate stress and provide higher yield in varying climatic conditions. Current achievements in genomics, proteomics, and metabolomics have offered remarkable opportunities to develop improved maize. This chapter highlights the necessity for increasing maize tolerance to drought and heat waves, review the best shoot and root characteristics and phenotypes, and propose an ideotype for sustainable maize production under fluctuating climatic conditions. This would help in improving maize by using molecular tools in a conventional breeding program (Gong et al., 2015).

A maize ideotype that utilizes maximum optimum production environment was developed. The optimum production environment factors comprise (a) sufficient moisture; (b) favorable temperatures all over the growing season; (c) satisfactory fertility; (d) high plant densities; (e) narrow row spacings; and (f) early dates of planting. The traits of the maize ideotype that provides optimum yield when culti-vated in such an environment are (a) stiff, vertically oriented leaves above the ear and horizontally oriented leaves below the ear; (b) high photosynthetic efficiency; (c) effective conversion of photosynthates to grain; (d) short interval between pollen shed and silk emergence; (e) prolificacy of ear-shoot; (f) small tassels; (g) photoperiod insen-sitivity; (h) cold-tolerance in germinating seeds and young seedlings; (i) long grain-filling period; and (j) slow leaf senescence (Mock and Pearce, 1975).

In grasses, the degradability of a cell wall is associated with the lignin content and the ferulic-mediated cross-linking of lignins to polysaccharides. Using image analyses 22 maize inbred lines were examined for the variations in degradability of cell wall. Then lignins and *p*-hydroxycinnamic acid contents were analyzed chemically and the proximity of biochemical and histological estimates of lignin levels was recognized for the first time. The combination of histological and biochemical characteristics explained only 89% of the variations for cell wall degradability and described a maize ideotype. It was found that the maize ideotype contains reduced lignin level and lignins were richer in syringyl than in guaiacyl units. This richness of syringyl units would support wall degradability in grasses and could be related to the fact that grass syringyl units are *p*-coumaroylated. This may influence the lignins and polysaccharides interaction abilities (Méchin et al., 2005).

Achieving sustained food production by enhancing N-use efficiency in intensive cropping systems is a major concern for scientists, environmental groups, and agricultural policymakers. In maize, nitrogen loss occurs in the form of nitrate through leaching. In this chapter, a maize ideotype with root architecture for efficient N uptake is discussed. The characteristics of ideotype include (i) deep roots that can exploit nitrate before it leaches down into deep soil; (ii) lateral root with vigorous growth under high N conditions in order to increase availability of spatial N in the soil; and (iii) strong response of lateral root growth to local nitrogen input to exploit unevenly distributed nitrate particularly under low N conditions (Mi et al., 2010).

As water and nitrate move in deeper soil strata with time and are initially get depleted in surface soil strata, a root system that can rapidly exploit the deep soil would increase water and N acquisition. Based on the idea that soil resource acquirement can be enhanced by the concurrence of root foraging and resource availability in time and space a maize hypothetical ideotype is presented. The characteristics *specific to ideotype* that affect the rooting depth in maize comprise *(a)* PR with large diameter, long LR and tolerance to low soil temperatures, *(b)* several seminal roots with small diameter, shallow growth angles, several LR, and long root hairs, (*c*) many CR with steep growth angles, and only some long laterals, (*d*) one whorl of BR with shallow growth angle and few long laterals, (*e*) abundant cortical aerenchyma, large cortical cell size, an optimum number of cells per cortical creating low cortical respiratory burden and faster cortical senescence, (*f*) unresponsiveness of lateral branching to local resource availability, and (*g*) low K_m and high V_{max} for the uptake of nitrate. Only a few characters of this ideotype have experimental support, others are assumed. Most of these characters of ideotype are pertinent to other cereal root systems and mostly to root systems of dicot crops. Further, owing to the significance of deep rooting for water uptake, this ideotype is pertinent to low-input systems (Lynch, 2013).

To reduce losses of maize yield under drought-prone environments, the improved maize varieties need to be developed by means of a multidisciplinary approach. The elucidation of important growth processes provides opportunities to develop maize lines tolerant to drought by identifying important phenotypic characters, ideotypes, and donors. Here, a set of tropical and subtropical maize inbredlines and single cross hybrids were tested at their reproductive stage under water deficit and well-watered conditions. The patterns of biomass production, senescence, and plant water status were observed all over the crop growth period. Under well-watered conditions, the stay-green pattern was crucial for hybrid yield, whereas the capability to uphold a high biomass all over the growth period was important for inbred yield. Under water deficit conditions, early biomass production before anthesis was essential for inbred yield, whereas delayed senescence was significant for hybrid yield. Further, the growth and senescence patterns were characterized using a new quantitative phenotyping tool, spectral reflectance (Normalized Difference Vegetation Index), and the qualitative measurements of canopy senescence were found to be related to grain yield (Cairns et al., 2012).

Though mixed cropping systems offer multiple ecosystem services, plant genotypes are hardly developed for these systems. In Rwanda, bean (*Phaseolus vulgaris* L.) and maize (*Zea mays* L.) intercrop is most commonly followed mixed cropping system. It was found that farmers consider diverse characteristics for different cropping systems. Farmers evaluate the intercrops based on five aspects: general characters and trait-based competitive ability, inherent competitive ability, adaptation to environment, and management. Farmers deliberate intrinsic competitive ability as an important factor while most other studies have neglected this characteristic in intercrop-breeding methods. Therefore, empowering farmers through on-farm testing of different genotypes would be a feasible solution to costly genotype evaluation by environment trials. This would help to identify highly adaptive and productive genotypes for diverse and resilient cropping systems (Isaacs et al., 2016).

4.2.1 IDEOTYPE TRAITS UNDER INTENSIVE CROPPING

Mi et al. (2016) proposed and described in detail two models of RSA for maize ideotype both at the seedling stage and also the adult stage. The root system of maize supports the growth of shoot, for nutrient and water uptake. In intensive planting systems, the maize population density is very high, wherein the nutrients are supplemented externally through intensive fertilizers to meet the requirements of maize plant growth and grain yields. Earlier researches have proposed ideotype characteristics of RSA for water and nutrient uptake and water deficit conditions. They have proposed this ideotype for high-intensity cropping systems for efficient resource utilization and efficiency. They mentioned that the RSA while proposing an ideotype should be able to meet the efficient usage of water and erratically distributed nutrients in the soil; concomitantly it should also possess larger root resistance for lodging. Additionally, it must be able to adapt for the resourceful use of the nutrients and carbon within the plant. In maize, both the embryonic root system and also the post embryonic root system are under separate genetic mechanisms. These two root systems also vary in their functions at varied soil environmental conditions as well as growth stages. They also described the root characteristic traits of this ideotype of both the axile and LR.

4.3 CONCLUSIONS

Crop ideotype should have plant features like leaf morphology, its orientation, branching patterns, canopy for effective capture of sunlight contributing to high production of the crops. Several concepts and features of ideotype were explained by many authors for maize. In this chapter, various research activities on maize plant architecture and maize ideotype are briefly reviewed.

KEYWORDS

- maize
- ideotype
- plant architecture
- characteristics

REFERENCES

Brown, P. J.; Upadyayula, N.; Mahone, G. S.; Tian, F.; Bradbury, P. J.; Myles, S.; Holland, J. B.; Flint-Garcia, S.; McMullen, M. D.; Buckler, E. S.; Rocheford, T. R. Distinct Genetic Architectures for Male and Female Inflorescence Traits of Maize. *PLoS Genet.* **2011**, *7*, e1002383.

Bouchet, S.; Bertin, P.; Presterl, T.; Jamin, P.; Coubriche, D.; Gouesnard, B.; Laborde, J.; Charcosset, A. Association Mapping for Phenology and Plant Architecture in Maize

Shows Higher Power for Developmental Traits Compared with Growth Influenced Traits. *Heredity* **2017**, *118*, 249–259.

Bowman, J.L.; Smyth, D. R. CRABS CLAW, A Gene that Regulates Carpel and Nectary Development in *Arabidopsis*, Encodes a Novel Protein with Zinc Finger and Helix-Loop-Helix Domains. Development **1999**, *126*, 2387–2396.

Cai, H.; Chen, F.; Mi, G.; Zhang, F.; Maurer, H. P.; Liu, W.; Reif, J. C.; Yuan, L. Mapping QTLs for Root System Architecture of Maize (*Zea mays* L.) in the Field at Different Developmental Stages. *Theor. Appl. Genet.* **2012**, *125*, 1313–1324.

Cairns, J. E.; Sanchez, C.; Vargas, M.; Ordoñez, R.; Araus, J. L. Dissecting Maize Productivity: Ideotypes Associated with Grain Yield under Drought Stress and Well-Watered Conditions. *J. Integr. Plant Biol.* **2012**, *54*, 1007–1020.

Dechorgnat, J.; Francis, K. L.; Dhugga, K. S.; Rafalski, J. A.; Tyerman, S. D.; Kaiser, B. N. Root Ideotype Influences Nitrogen Transport and Assimilation in Maize. *Front. Plant Sci.* **2018**, *9*, 531.

Dong, L.; Qin, L.; Dai, X.; Ding, Z.; Bi, R.; Liu, P.; Chen, Y.; Brutnell, T. P.; Wang, X.; Li, P. Transcriptomic Analysis of Leaf Sheath Maturation in Maize. *Int. J. Mol. Sci.* **2019**, *20*(10), 2472. https://doi.org/10.3390/ijms20102472.

Dong, Z.; Li, W.; Unger-Wallace, E.; Yang, J.; Vollbrecht, E.; Chuck, G. Ideal Crop Plant Architecture Is Mediated by Tassels Replace Upper Ears1, a BTB/POZ Ankyrin Repeat Gene Directly Targeted by TEOSINTE BRANCHED1. *Proc. Natl. Acad. Sci.* **2017**, *114*, E8656–E8664.

Feng, T.; Bradbury, P. J.; Brown, P. J.; Hung, H.; Sun, Q.; Flint Garcia, S.; Rocheford, T. R.; McMullen, M. D.; Holland, J. B.; Buckler, E. S. Genome-Wide Association Study of Leaf Architecture in the Maize Nested Association Mapping Population. *Nat. Genet.* **2011**, *43*, 159–162.

Giuliani, S.; Sanguineti, M. C.; Tuberosa, R.; Bellotti, M.; Salvi, S.; Landi, P. Root-ABA1, a Major Constitutive QTL, Affects Maize Root Architecture and Leaf ABA Concentration at Different Water Regimes. *J. Exp. Bot.* **2005**, *56*, 3061–3070.

Gong, F.; Wu, X.; Zhang, H.; Chen, Y.; Wang, W. Making Better Maize Plants for Sustainable Grain Production in a Changing Climate. *Front. Plant Sci.* **2015**, *6*, 835.

Hammer, G. L.; Dong, Z.; McLean; G.; Doherty, A.; Messina C.; Schussler, J.; Zinselmeier, C.; Paskiewicz, S.; Cooper, M. Can Changes in Canopy and/or Root System Architecture Explain Historical Maize Yield Trends in the U.S. Corn Belt? *Crop Sci.* **2009**, *49*, 299–312.

Hochholdinger, F.; Tuberosa, R. Genetic and Genomic Dissection of Maize Root Development and Architecture. *Curr. Opin. Plant Biol.* **2009**, *12*, 172–177.

Hochholdinger, F.; Yu, P.; Marcon, C. Genetic Control of Root System Development in Maize. *Trends Plant Sci.* **2018a**, *23*, 79–88.

Hochholdinger, F.; Marcon, C.; Baldauf, J. A.; Yu, P.; Frey, F. P. Proteomics of Maize Root Development. *Front. Plant Sci.* **2018b**, *9*, 143. doi: 10.3389/fpls.2018.00143.

Isaacs, K. B.; Snapp, S. S.; Kelly, J. D.; Chung, K. R. Farmer Knowledge Identifies a Competitive Bean Ideotype for Maize–Bean Intercrop Systems in Rwanda. *Agric. Food Security* **2016**, *5*, 1–15.

Johnston, R.; Sun, Q.; Sylvester, A. W.; Hake, S.; Scanlon, M. J. Transcriptomic Analyses Indicate that Maize Ligule Development Recapitulates Gene Expression Patterns that Occur During Lateral Organ Initiation. Plant Cell **2014**, *26*, 4718–4732.

Juarez, M. T.; Twigg, R. W.; Timmermans, M. C. P. Specification of Adaxial Cell Fate During Maize Leaf Development. *Development* **2004,** *131,* 4533–4544.

Khakwani, K.; Rafique, M.; Malhi, A. R; Altaf, M.; Saleem, S.; Arshad, M. Maize Ideotype Breeding for Changing Environmental Conditions. *Afr. J. Agric. Res.* **2018,** *13,* 512–517.

Ku, L. X.; Zhao, W. M.; Zhang, J.; Wu, L. C.; Wang, C. L.; Wang, P. A.; Chen, Y. H. Quantitative Trait Loci Mapping of Leaf Angle and Leaf Orientation Value in Maize (*Zea mays* L.). *Theor. Appl. Genet.* **2010,** *121,* 951–959.

Lewis, M. W.; Bolduc, N.; Hake, K.; Htike, Y.; Hay, A.; Candela, H.; Hake, S. Gene Regulatory Interactions at Lateral Organ Boundaries in Maize. *Development* **2014,** *141,* 4590–4597.

Li, Z.; Zhang, X.; Zhao, Y.; Li, Y.; Zhang, G.; Peng, Z.; Zhang, J. Enhancing Auxin Accumulation in Maize Root Tips Improves Root Growth and Dwarfs Plant Height. *Plant Biotechnol. J.* **2018,** *16,* 86–99.

Li, C.; Li, Y.; Shi, Y.; Song, Y.; Zhang, D.; Buckler, E. S.; Li, Y. Genetic Control of the Leaf Angle and Leaf Orientation Value as Revealed by Ultra-High Density Maps in Three Connected Maize Populations. *PLoS One* **2015,** *10,* e0121624.

Li, D.; Wang, X.; Zhang, X.; Chen, Q.; Xu, G.; Xu, D.; Wang, C.; Liang, Y.; Wu, L.; Huang, C.; Tian, J.; Wu, Y.; Tian, F. The Genetic Architecture of Leaf Number and Its Genetic Relationship to Flowering Time in Maize. *New Phytol.* **2016,** *210,* 256–268.

Lockhart, J. Exploring Maize Leaf Architecture from Different Angles. *Plant Cell* **2017,** *29,* 1550–1551.

Lynch, J. P. Steep, Cheap, and Deep: An Ideotype to Optimize Water and N Acquisition by Maize Root Systems. *Ann. Bot.* **2013,** *112,* 347–357.

Maddonni, G. A.; Otegui, M. E.; Cirilo, A. G. Plant Population Density, Row Spacing and Hybrid Effects on Maize Canopy Architecture and Light Attenuation. *Field Crops Res.* **2001,** *71,* 183–193.

Martínez-de la Cruz, E.; García-Ramírez, E.; Vázquez-Ramos, J. M.; Reyes de la Cruz, H.; López-Bucio, J. Auxins Differentially Regulate Root System Architecture and Cell Cycle Protein Levels in Maize Seedlings. *J. Plant Physiol.* **2015,** *176,* 147–156.

Mejía-Guerra, M. K.; Buckler, E. S. Ak-mer Grammar Analysis to Uncover Maize Regulatory Architecture. *BMC Plant Biol.* **2019,** *19,* 1–17.

Mi, G.; Chen, F.; Yuan, L.; Zhang, F. Ideotype Root System Architecture for Maize to Achieve High Yield and Resource Use Efficiency in Intensive Cropping Systems. *Adv. Agron.* **2016,** *139,* 73–97.

Mi, G. H.; Chen, F. J.; Wu, Q. P.; Lai, N. W.; Yuan, L. X.; Zhang, F. S. Ideotype Root Architecture for Efficient Nitrogen Acquisition by Maize in Intensive Cropping Systems. *Sci. China Life Sci.* **2010,** *53,* 1369–1373.

Mock, J. J.; Pearce, R. B. An Ideotype of Maize. *Euphytica* **1975,** *24,* 613–623.

Pace, J.; Gardner, C.; Romay, C.; Ganapathysubramanian, B.; Lübberstedt, T. Genome-Wide Association Analysis of Seedling Root Development in Maize (Zea mays L.). *BMC Genomics* **2015,** *16*(1), 47. https://doi.org/10.1186/s12864-015-1226-9.

Pautler, M.; Eveland, A. L.; LaRue, T.; Yang, F.; Weeks, R.; Lunde, C.; Je, B.; Meeley, R.; Komatsu, M.; Vollbrecht, E.; Sakai, H.; Jackson, D. Fasciated EAR4 Encodes a bZIP Transcription Factor that Regulates Shoot Meristem Size in Maize. *Plant Cell* **2015,** *27,* 104–120.

Peiffer, J. A.; Romay, M. C.; Gore, M. A.; Flint-Garcia, S. A.; Zhang, Z.; Millard, M. J.; Gardner, C. A. C.; McMullen, M. D.; Holland, J. B.; Bradbury, P. J.; Buckler, E. S. The Genetic Architecture of Maize Height. *Genetics* **2014**, *196*, 1337–1356.

Perez, R. P. A.; Fournier, C.; Cabrera-Bosquet, L.; Artzet, S.; Pradal, C.; Brichet, N.; Chen, T. W.; Chapuis, R.; Welcker, C.; Tardieu, F. Changes in the Vertical Distribution of Leaf Area Enhanced Light Interception Efficiency in Maize Over Generations of Selection. *Plant, Cell Environ.* **2019**, *42*, 2105–2119. doi:10.1111/pce.13539.

Postma, J. A.; Dathe, A.; Lynch, J. P. The Optimal Lateral Root Branching Density for Maize Depends on Nitrogen and Phosphorus Availability. *Plant Physiol.* **2014**, *166*, 590–602.

Salvi, S.; Giuliani, S.; Ricciolini, C.; Carraro, N.; Maccaferri, M.; Presterl, T.; Ouzunova, M.; Tuberosa, R. Two Major Quantitative Trait Loci Controlling the Number of Seminal Roots in Maize Co-map with the Root Developmental Genes RTCS and RUM1. *J. Exp. Bot.* **2016**, *67*, 1149–1159.

Singh, V.; van Oosterom, E. J.; Jordan, D. R.; Messina, C. D.; Cooper, M.; Hammer, G. L. Morphological and Architectural Development of Root Systems in Sorghum and Maize. *Plant Soil* **2010**, *333*, 287–299.

Strable, J.; Wallace, J. G.; Unger-Wallace, E.; Briggs, S.; Bradbury, P. J.; Buckler, E. S.; Vollbrecht, E. Maize YABBY Genes Drooping Leaf1 and Drooping Leaf2 Regulate Plant Architecture. *Plant Cell* **2017**, *29*, 1622–1641.

Tian, J.; Wang, C.; Xia, J.; Wu, L.; Xu, G.; Li, D.; Qing, W.; Han, X.; Chen, Q.; Jin, W.; Tian, F. Teosinte Ligule Allele Narrows Plant Architecture and Enhances High-Density Maize Yields. *Science* **2019**, *365*, 658–664.

Thompson, A. M.; Yu, J.; Timmermans, M. C. P.; Schnable, P.; Crants, J. E.; Scanlon, M. J.; Muehlbauer, G. J. Diversity of Maize Shoot Apical Meristem Architecture and Its Relationship to Plant Morphology. *Genes/Genomes Genet.* **2015**, *5*, 819–827.

Trachsel, S.; Kaeppler, S. M.; Brown, K. M.; Lynch, J. P. Shovelomics: High Throughput Phenotyping of Maize (*Zea mays* L.) Root Architecture in the Field. *Plant Soil* **2011**, *341*, 75–87.

Méchin, V.; Argillier, O.; Rocher, F.; Hébert, Y.; Mila, I.; Pollet, B.; Barriére, Y.; Lapierre, C. In Search of a Maize Ideotype for Cell Wall Enzymatic Degradability Using Histological and Biochemical Lignin Characterization. *J. Agric. Food Chem.* **2005**, *53*, 5872–5881.

Wei, L.; Zhang, X.; Zhang, Z.; Liu, H.; Lin, Z. A New Allele of the Brachytic2 Gene in Maize Can Efficiently Modify Plant Architecture. *Heredity* **2018a**, *121*, 75–86.

Wei, H.; Zhao, Y.; Xie, Y.; Wang, H. Exploiting SPL Genes to Improve Maize Plant Architecture Tailored for High-Density Planting. *J. Exp. Bot.* **2018b**, *69*, 4675–4688.

Wei, X.; Wang, X.; Guo, S.; Zhou, J.; Shi, Y.; Wang, H.; Chen, Y. Epistatic and QTL × Environment Interaction Effects on Leaf Area - Associated Traits in Maize. *Plant Breed.* **2016**, *135*, 671–676.

Xu, G.; Wang, X.; Huang, C.; Xu, D.; Li, D.; Tian, J.; Chen, Q.; Wang, C.; Liang, Y.; Wu, Y.; Yang, X.; Tian, F. Complex Genetic Architecture Underlies Maize Tassel Domestication. *New Phytol.* **2017**, *214*, 852–864.

York, L. M.; Lynch, J. P. Intensive Field Phenotyping of Maize (*Zea mays* L.) Root Crowns Identifies Phenes and Phene Integration Associated with Plant Growth and Nitrogen Acquisition. *J. Exp. Bot.* **2015**, *66*, 5493–5505.

Zhang, D.; Sun, W.; Singh, R.; Zheng, Y.; Cao, Z. GRF-Interacting Factor1 Regulates Shoot Architecture and Meristem Determinacy in Maize. *Plant Cell* **2018a**, *30*, 360–374.

Zhang, Z.; Zhang, X.; Lin, Z.; Wang, J.; Xu, M.; Lai, J.; Yu, J.; Lin, Z. The Genetic Architecture of Nodal Root Number in Maize. *Plant J.* **2018b,** *93*, 1032–1044.

Zhang, J.; Ku, L. X.; Han, Z. P.; Guo, S. L.; Liu, H. J.; Zhang, Z. Z.; Cao, L. R.; Cui, X. J.; Chen, Y. H. The *ZmCLA4* Gene in the qLA4-1 QTL Controls Leaf Angle in Maize (*Zea mays* L.). *J. Exp. Bot.* **2014,** *65*, 5063–5076.

CHAPTER 5

Maize Botany

ABSTRACT

In this chapter, a brief overview of different aspects of botany such as taxonomy, morphology, anatomy of roots, stems, their characterization and classification, and relation to adaptation to abiotic stresses such as drought and flood is presented. The molecular characterization of these aspects is also discussed.

5.1 TAXONOMY

In this study, the utility of fatty acid composition as an index of maize population classification was discussed. A total of 102 maize populations were collected from the Italian germplasm and classified based on their grain fatty acid contents. Palmitic acid content was negatively correlated with linoleic acid content. The progressive increase in linoleic acid contents was found to be related with increasing lateness and adaptation to hotter climates. But the morphological characteristics do not show any correlation with the fatty acid contents. So, this new classification method can be applied to improve classifications obtained by the analysis of morphological traits (Camussi et al., 1980).

Louette et al. (1997) discussed the significance of models developed to study the *in situ* conservation of crop genetic resources on the basis of the geographical separation of a community. It was found that the introduction of both improved cultivars and landraces from other communities enhanced the morphophenological diversity of local materials. Alternatively, for defining "local" landrace, the geographical direction was shown to be larger than the community itself. Generally, farmers use phenotypic characteristics to differentiate varieties and categorize seeds taken from other farmers in and outside the community. Therefore, it can be concluded that the maize diversity in this community was a result of genetic material introduction and not of geographical isolation.

Bellon and Brush (1994) investigated the domestication center of maize viz. Mesoamerica for its biological diversity. It was found that in central Chiapas of Southern México maize farmers cultivate only 15 local varieties belonging to six races and four race mixtures. Though the modern and high yielding varieties of maize were adopted in wide areas, the maize farmer's select only local varieties for specific soils because of agronomic and use standards. As the spatial and temporal separation seems to be insufficient to maintain varieties, farmers maintain them mainly through seed selection. The hybridization between local maize populations can lead to the formation of a heterogeneous population from uniform a population of improved varieties.

In maize, a high level of genetic diversity has been identified. Gore et al. (2009) assumed a study to give an insight into maize genetic diversity. A group of 27 diverse maize inbred lines were genotyped and millions of sequence polymorphisms were detected. A high level of divergent haplotypes and 10- to 30-fold variation in recombination rates were observed. It was found that most of the chromosomes in maize are pericentromic type with highly suppressed recombination. These suppressed regions may form a major part of heterosis, influencing the efficiency of selection during the development of maize inbreds. Additionally, hundreds of selective sweeps and highly differentiated regions holding loci crucial for geographic adaptation were identified. Thus, this study has provided a base for joining breeding efforts throughout the world and for studying complex traits by means of genome-wide association studies.

This experiment was planned to improve the classification of the Mexican races of maize as part of the process of revising the Razas de Maiz en Mexico. The inter-relationships among the races are studied by numerical taxonomy of morphological characters and comparing classifications with earlier studies. Forty-nine Mexican races, denoted by 148 collections, were cultivated in several locations and seasons in Mexico from 1982 to 1984; 47 characters were assessed directly. For the analysis using numerical taxonomy, characters with the ratio ($r=[\sigma^2r/(\sigma^2re+\sigma^2e)] \geq 3.0$) were selected. Classifications of Mexican races indicated general conformity with the associations observed in previous studies that were based on conventional taxonomic methods and numerical taxonomy. Furthermore, poorly defined races and novel types may now be allocated to well defined groups (Sanchez and Goodman, 1992).

A compromise classification of the genus *Zea*, revealing both phylogeny and practical needs, identifies six taxa, as follows: Section LUXURI-ANTES: *Zea perennis, Zea diploperennis, Zea luxurians*. Section ZEA: *Zea mays* ssp. mexicana (Neo-volcanic Plateau), *Zea mays ssp. parviglumis, Zea mexicana ssp. parviglumis* var. huehuetenangensis, *Zea mays ssp. mays*. The new subspecies is differentiated by smaller spikelets and rachis joints, the varieties by different habitats, blooming dates, and their genetic behavior with regard to cultivated *Zea mays*. *Zea mays ssp. mexicana* is the ancestor of corn. The cultivated "CORN" OR "MAIZE" of the American Indians and world commerce, *Zea mays ssp. mays* is a species whose taxonomic limitation has never been questioned. Its wild relatives, the "teosintes," however, have for quite some time now been considered by many workers to present a number of taxonomic complications (Iltis and Doebley, 1980).

Molecular DNA analyses of the New World grass (Poaceae) genus *Zea*, including five species, has sort out taxonomic issues comprising the most probable teosinte progenitor (*Zea mays* ssp. parviglumis) of maize (*Zea mays* ssp. mays). But, archaeologically, much information is not available about the usage of teosinte by humans both before and after the maize domestication. Opaline phytoliths produced in teosinte fruit cases are a probable line of indication to study these associations. In this study, multi-dimensional scaling and multiple discriminant analyses were used to find out if rondel phytolith assemblages from teosinte fruit cases reveal teosinte taxonomy. The findings indicated that rondel phytolith assemblages from the different taxa, including subspecies, can be statistically differentiated. This shows that it will be possible to explore the archaeological histories of teosinte (Hart et al., 2011).

Plant genetic diversity is the most important part of any agricultural ecosystem. Therefore, it is essential to classify genetic resources properly to conserve, assess, and improve germplasm effectively. In maize (*Zea mays* L.), many classification systems have been used for defining maize races. From the 1980s, with the application of computers, numerical taxonomy became increasingly significant and multivariate methods began to be used for categorizing genetic resources. Gutiérrez et al. (2003) compared two methods of classification of Uruguayan maize landraces viz., an earliest racial classification acquired through visual assessment and a numerical classification. A two-stage classification strategy was used to conduct numerical classification. First, primary groups were made by the Ward method and, then, the modified location model (MLM) refined those groups. This classification was compared with the earliest racial classification by four criteria. The Ward-MLM approach produced more homogeneous groups than those analogous to the earliest racial classification. The numerical classification conserved the structure of the more diverse races but separated the Cateto Sulino race into two more homogeneous groups, each with smaller variance and more diverse than other groups. Groups with clearly distinct features, in terms of the numerical variables, and better, in terms of the four criteria used were produced in numerical classification. Therefore, these results would form a basis to improve the racial classification of maize landraces of Uruguay.

5.2 MORPHOLOGY

The cotyledon in maize, oats, and wheat is known as scutellum. Sometimes, the "ventral scale" of the cotyledon is interpreted as its ligule. The epiblast, which probably has little morphological importance, can not be deliberated a rudimentary cotyledon. The maize coleoptile viz, second leaf of the plant is homologous with a foliage leaf and the leaf distal to it is the third leaf of the plant. In maize and oat seedlings, the elongated structure between the cotyledon and the coleoptile is the first internode of the axis, and the term "mesocotyl" used for this structure is irrelevant (Fig. 5.1). In the young wheat axis, the main elongating structure is the second internode, which is always enclosed by the coleoptile. In grasses, the hypocotyledonary region of "transition" is absent. The transition begins and takes place mainly in the vascular plate at the first node, and in the first internode. The transition persists in the second internode and to a certain extent in the third and fourth internodes also. These morphological features (Fig. 5.2) were often observed in mature embryos, before the beginning of germination. In addition to this, the vascular skeleton similarities have been pointed out in this review (Avery, 1930).

Eighty six open-pollinated varieties of maize were collected from an extensive climatic and geographical spectrum of Spain and classified based on their morphological traits. Using multivariate analysis methods such as principal component and Cluster analysis, data of 23 characters associated with earliness, plant and tassel structure, and the shape of the ear and grain were analyzed. About 64.5% of the total variation was represented by the first three principal components, which indicates the relative discriminant value of each of the traits under consideration. Then, using

Mahalanobis' generalized distance, the populations were grouped into four distinct subracial groups based on the degree of their dissimilarity. The groups were related to four geographically or climatically demarcated areas. Therefore, this study suggested a simple hierarchical system of classification based on the morphological characters in maize (Llaurado and Moreno Gonzalez, 1993).

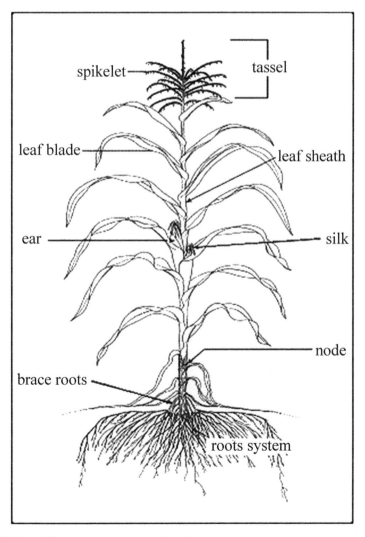

FIGURE 5.1   Diagrammatic representation of maize plant.

FIGURE 5.2   Maize plant showing aerial parts.

A study has been carried out by Sanchez et al. (2000) to determine the association and genetic diversity among the 209 accessions of maize obtained from 59 Mexican races. Out of the 209 accessions, 154 accessions were cultivated in multiple locations and seasons in Mexico and examined for 21 enzyme systems encoded by 37 loci. The remaining 47 accessions were evaluated for morphological characters. The results revealed high genetic variation among and within the Mexican races. In these accessions, about 65% of the alleles were found to be rare and occurred at frequencies below 0.01. Furthermore, some accessions exhibited low genetic diversity and high genetic differentiation resembling self-pollinated crops. Most of the accessions with genetic diversity values were field varieties.

Mixed cropping is followed extensively in developing countries and is receiving increasing interest for sustainable agriculture in developed countries. Plants in intercrops grow in a different way from plants in single crops, because of interspecific plant interactions, but adaptive plant morphological responses to competition in mixed stands have not been considered in detail. Here, the maize (*Zea mays*) response to mixed cultivation with wheat (*Triticum aestivum*) is explained. Findings revealed that early responses of maize to the altered light environment in mixed stands spread throughout maize development, resulting in diverse phenotypes compared with pure stands. They compared photosynthetically active radiation (PAR), red:far-red ratio (R:FR), leaf development, and final organ sizes of maize grown in three cultivation systems: pure maize, an intercrop with a small distance (25 cm) between maize and wheat plants, and an intercop with a large distance (44 cm) between the maize and the wheat. Compared with maize in pure stands, maize in the mixed stands had lower leaf and collar appearance rates, increased blade and sheath lengths at low ranks and smaller sizes at high ranks, increased blade elongation duration, and reduced R: FR and PAR at the plant base during early development. The treatment with a short distance between wheat and maize strips showed greater effects. The results suggested a reaction between leaf initiation and leaf emergence at the plant level and coordination between blade and sheath growth at the phytomer level. A theoretical model, based on coordination rules, was projected to elucidate the development of the maize plant in pure and mixed stands (Zhu et al., 2014).

5.3 ANATOMY

Blackman and Davies (1985) considered the effects of soil drying on maize root and shoot systems. The maize seedlings were grown by dividing their roots between two containers, such that the root part in one container was supplied with water, while the root part in the other container was exposed to reduced soil water potential. The plants in half watered container showed partial closure of stomata; however, leaf water potential, turgor, and abscisic acid (ABA) content remained unaffected. When the leaves from these half watered plants were detached and incubated under conditions favorable for stomatal opening, the plants still exhibited limited stomatal apertures. Afterward, the closure of stomata accelerated by soil drying was reversed by incubating the plants in kinetin (10 mmol m^{-3}) or zeatin (100 mmol m^{-3}). These findings suggested that a constant supply of cytokinin from roots may be required to support maximum stomata opening, and stoppage of this supply due to soil drying indicates the inhibition of root activity, leading to restricted stomatal opening and thus limited usage of water.

5.4   MORPHOMETRIC VARIATION IN SHOOT APICAL MERISTEM

In the shoot apical meristem, a miniature group of cells termed as stem cells are involved in the generation of the above ground organs. These cells are also common in the apical meristems of maize shoot. Though, several mutational studies have recognized the genetic networks involved in the regulation of functions of shoot apical meristem, very little is known on the morphological variations in the natural maize populations. Leiboff et al. (2015) reported about the use of a high-throughput image processing for capturing this variation in size of the shoot apical meristems in diverse maize inbred panel plants. They have demonstrated that there was the presence of correlations between the shoot apical meristem size in a seedling and that of agronomically significant adult characteristics such as flowering time, stem size and leaf node number, etc. Genome wide association study which involved a combination of shoot apical meristem phenotypes with 1.2 million single-nucleotide polymorphisms (SNPs) has shown the presence of unexpected shoot apical meristem morphology candidate genes. Analyses of these candidate genes implicated in hormone transport, cell division and cell size prove correlations between SAM morphology and trait-associated SNP alleles. Their data has illustrated that a microscopic shoot apical meristem of a seedling is extrapolative of adult phenotypes. Further, they mentioned that morphometric dissimilarities in the shoot apical meristem are mainly associated with genes which in the earlier studies were not predicted to be involved in regulating the size of the shoot apical meristem in maize.

Several greenhouse or growth chamber experiments have demonstrated that pots can have a limiting effect on plant growth. So as to study the effects of pot size on maize plants especially on transpiration response under water-deficit stress, Ray and Sinclair (1998) undertook an experiment. They have grown the maize plants in 2.3, 4.1, 9.1, and 16.2 pots and to avoid water loss except by transpiration pots were sealed. Plants in each pot size were divided into two watering systems, a well-watered control and a water-deficit treatment. Water deficits were inflicted by simply not re watering the pots. With the decreasing pot size, a substantial reduction in shoot dry weight and total transpiration were observed under both watering regimes. But, the plants in different pot size did not show significant differences in the fraction of transpirable soil water (*FTSW*) point at which transpiration start declining (*FTSW* ≈ 0.31 for maize) or in the association of transpiration rate to soil water content in response to water deficits. Thus the study findings revealed that the soil water content is the key factor regulating transpirational response to drought stress irrespective of pot size or plant size.

Many studies have shown that silicon (Si) ameliorates the negative effects of cadmium (Cd) on the growth and development of plants. But the mechanism of this phenomenon is not fully recognized. In this regard, Vaculík et al. (2012) explained the effect of Si on plant growth on Cd uptake and subcellular distribution in maize plants relative to the development of root tissue. Young maize seedlings (*Zea mays*) were grown hydroponically with 5 or 50 μM Cd and/or 5 mM Si for 10 days. Then with atomic absorption spectrometry or inductively coupled plasma mass spectroscopy growth parameters and the concentrations of both Cd and Si were measured in root and shoot. The apoplasmic barriers and vascular tissue development in roots and the effect of Si on apoplasmic and symplasmic distribution of [109]Cd applied at 34 nM was examined among root and shoot.

Previous studies reported that the sunlight regulates the formation of anthocyanin pigment in vegetative tissues of maize and noticed only in the cultivars containing homozygous recessive *pl* loci. But the clear evidences on the photoreceptor regulated production of anthocyanin pigment are not available. Here, the nature of photoreceptor(s) facilitating this anthocyanin formation was analyzed in maize hybrid, Kanchan-521.The anthocyanin pigment formation was seen in all the vegetative organs of maize seedlings exposed to sunlight. After exposing the seedlings to sunlight the slow increase in photoinduction of anthocyanin was observed between 4–16 h and a rapid increment was noted between 16–24 h. They found that the UV-B component of sunlight primarily mediates photoinduction of anthocyanin, which can be elicited by exposing the seedlings to an artificial UV-B light source. When the sunlight exposure was stopped with a far-red pulse before transferring the seedlings to darkness, a decrease in the sunlight-regulated induction of anthocyanin was observed. This indicates a coaction of phytochrome in this photoresponse. Further, the exposure to sunlight was found to stimulate the activity of phenylalanine ammonia lyase (PAL) in all organs with two temporally parted peaks. One peak of PAL between 4 and 12 h was stimulated by phytochrome, and the other peak between 12 and 24 h was induced by UV-B light. The findings of this study indicated that UV-B light and phytochrome together regulate the photoinduction of anthocyanin in maize (Singh et al., 1999).

Maize is a model crop and the information on genome sequence contributing to particular plant phenotypes and on the temporal and spatial transcription patterns of genes is essential. In this study, a complete atlas of global transcription profiles through growth stages and plant organs was presented. Transcription patterns in 60 different tissues from 11 major organ systems of inbred line B73 were profiled. They used NimbleGen microarray comprising 80,301 probe sets for profiling. Out of the 30,892 probe sets representing the filtered B73 gene models, 91.4% were detected in at least one tissue. In all, the tissues 44.5% of the probe sets were expressed. This indicates an extensive overlap of gene expression among plant organs. Then, biologically related tissues were grouped on the basis of these gene expression profiles. Further, they used these data to analyze the expression of genes encoding enzymes in the lignin biosynthetic pathway and located that the expansion of different gene families was associated with the divergent, tissue-specific transcription patterns of the paralogs. This atlas could be effectively utilized in the identification of genes and functional characterization of maize (Sekhon et al., 2011).

Auxin, a plant growth hormone, stimulates the initiation of lateral organs and meristems. In this study, a maize mutant *sparse inflorescence1* (*spi1*) with defective initiation of axillary meristems and lateral organs during the vegetative and inflorescence development stages was identified and characterized. Positional cloning has shown that the maize contains *spi1* gene that encodes a flavin monooxygenase, and it is comparable to the *YUCCA* (*YUC*) genes associated with the auxin biosynthesis in *Arabidopsis* plant tissues. Then, the different members of the *YUC* family in moss, monocot, and eudicot species were subjected to phylogenetic analysis. The phylogenetic analysis revealed that the *YUC* family expands independently in monocots and eudicots, and the role of individual *YUC* genes is differentiated within a monocot-specific clade, *containing spi1*. The expression and functional data suggested that *spi1* plays a dominant role in auxin biosynthesis that is crucial for the normal development of maize inflorescence. Additionally, the interaction analysis between *spi1* and genes regulating auxin transport indicated that auxin transport and biosynthesis work synergistically to control the development of axillary meristems and lateral organs in maize (Gallavotti et al., 2008).

Primary agriculturalists and modern farmers target maize architecture for selection because it affects harvesting, breeding methods, and mechanization. It was reported that the small groups of stem cells formed during vegetative and reproductive development viz., the activity of lateral meristems determine plant architecture. Then, the lateral meristems produce branches and inflorescence structures, which describe the general form of a plant. This study gives an account of the *barren stalk1* gene, isolation that encodes a noncanonical basic helix–loop–helix protein required for the initiation of all aerial lateral meristems in maize. It was observed that the *barren stalk1* gene and *teosinte branched1* gene together regulate the development of vegetative lateral meristem and patterning of maize inflorescences. They undertook sampling of nucleotide diversity in the *barren stalk1* region and found that from the wild progenitor of maize, teosinte, two haplotypes were passed into the maize gene pool and only one was integrated in modern inbreds. The study results indicated that *barren stalk1* was chosen for agronomic purposes (Gallavotti et al., 2004).

Maize stover is a main source of crop remains and a valuable sustainable energy in the United States. Stalk is the major constituent of stover and constitutes half of the stover dry weight. Therefore, to improve maize stover as a biofuel feedstock, the genetic factors of stalk traits need to be characterized. In this investigation, the candidate genes related with characters linked to stalk biomass and stalk anatomy were detected using genome-wide association studies (GWAS). The results unveiled 16 candidate genes related with four stalk traits. Most of these candidate genes were found to be engaged in basic cellular functions, such as gene expression regulation and progression of the cell cycle. Previous studies reported the association of regulatory genes such as Zmm22 and an ortholog of Fpa with plant height, which was proved in this study by means of a transgenic approach. Transgenic plants with increased expression of Zmm22 exhibited a substantial reduction in plant height and number of tassel branches, representing a pleiotropic effect of Zmm22. Genome-wide association analyses identified some candidate genes related with multiple traits, indicating common regulatory factors bring about different stalk traits. Thus, this paper has given an understanding of maize stalk anatomy and biomass genetic regulation that could be utilized to improve valuable stalk traits in breeding programs (Mazaheri et al., 2019).

Salt and drought stresses are major constraints in maize production (Hoque et al., 2018). In this study, the anatomical bases of salt and drought stress resistance were explored. They subjected 14-day-old seedlings of three maize hybrid genotypes to salt (100 mM NaCl) and drought stress (equiosmotic PEG6000) under hydroponic conditions. The salt and drought stresses do not show any effect on root protoxylem and metaxylem thickness and root diameter in drought resistant genotypes, while sensi-tive genotype (BARI hybrid maize-7) showed increased root protoxylem thickness. Under both salt and drought stresses, BARI hybrid maize12 exhibited unaffected response in leaf epidermal thickness, phloem area, xylem area, total leaf thickness, and increased bundle sheath thickness.

The homologous transcription factors FLORICAULA in *Antirrhinum* and LEAFY in *Arabidopsis* regulate flower meristem identity and its floral patterning. Though the functions of *FLORICAULA/LEAFY* homologs in the development of flower have been established in many dicots, their roles in more distantly related flowering plants, the monocots are yet to be known. Here, the role of two duplicate *FLORICAULA/LEAFY* homologs was studied in maize using reverse genetics. Transposons were inserted into the maize genes, *zfl1* and *zfl2, which caused a* disruption of floral organ identity and patterning, besides defects in inflorescence architecture and the vegetative to reproductive transition phase. The *zfl1*; *zfl2* double mutants phenotype suggested that the maize *FLORICAULA/LEAFY* homologs work as upstream regulators of the ABC floral organ identity genes. These results together with earlier reports indicated that the transcriptional network controlling the development of flower is partly conserved between monocots and dicots. Further, the study suggested that the *zfl* genes may perform a new role in regulating quantitative characteristics of inflorescence phyllotaxy in maize and can be applied for detecting the quantitative trait loci that control variances in inflorescence structure between maize and its progenitor, teosinte (Bomblies et al., 2003).

In apomixis, plants reproduce asexually through seeds without meiosis and fertilization. Though apomixis is regulated genetically, its essential genetic components have not been determined to date. Here, the sexual development in maize was compared to the apomixis in maize-*Tripsacum* hybrids with the help of profiling experiments. In ovules of apomictic plants, six specifically downregulated loci were identified. Out of six loci, four loci were homologous with the members of the RNA-directed DNA methylation pathway, which causes silencing through DNA methylation *in Arabidopsis thaliana*. Then, two maize DNA methyltransferase genes from the subset, *dmt102* and *dmt103*, were analyzed for their loss-of-function alleles. These genes were downregulated in the ovules of apomictic plants and were homologous to the *CHROMOMETHYLASEs* and *DOMAINS REARRANGED METHYLTRANSFERASE* families of *Arabidopsis*. The analysis unveiled the formation of unreduced gametes and multiple embryo sacs in the ovule. The expression of *dmt102* and *dmt103* genes was observed in a restricted domain and around the germ cells in the ovule. This suggested that an active DNA methylation pathway during reproduction is vital for the gametophyte development in maize and may perform a significant role in differentiating apomictic and sexual reproduction (Garcia-Aguilar et al., 2010).

To ascertain the secondary traits related with the heat, drought and combined drought and heat stress at the seeding stage of maize, a study has been conducted by Tandzi et al. (2019). Twenty genotypes were assessed by imposing stresses in a growth chamber and data on traits like leaf stress response percentage, leaf area, plant height, plant feature, and some indices (STI, HTI, DTI, and MSTI) were recorded. Under stress environments, major differences were noticed between genotypes for all characters. Shoot weight, plant height, and chlorophyll content exhibited a significant and positive association with stress tolerance indices (STI, HTI, DTI, and MSTI). This indicated that these traits could be considered while screening maize seedlings under combined drought and heat stress environments. Further, a strong and positive correlation was found between heat stress environment and combined drought and heat stress environment. Comparatively, a good performance was exhibited by three inbred lines L6-Y, L24-Y and Sweety 015 across environments and could be potential genotypes in breeding maize for tolerance to both drought and heat stress. Therefore, these findings demonstrated that the genotypes tolerant to heat stress probably tolerate combined drought and heat stress conditions.

5.5 CONCLUSIONS

Many researchers have been assumed on different aspects of botany such as taxonomy, morphology, anatomy of root, and stem, which were associated with adaptation to some abiotic stresses such as drought, flooding, etc. These research advances are briefly presented in this chapter.

KEYWORDS

- maize
- botany
- anatomy
- taxonomy
- molecular characterization

REFERENCES

Avery, G. S. Comparative Anatomy and Morphology of Embryos and Seedlings of Maize, Oats, and Wheat. *Int. J. Plant Sci.* **1930,** *89,* 1–39.

Bellon, M. R; Brush, S. B. Keepers of Maize in Chiapas, Mexico. *Econ. Bot.* **1994,** *48,* 196–209.

Blackman, P. G.; Davies, W. J. Root to Shoot Communication in Maize Plants of the Effects of Soil Drying. *J. Exp. Bot.* **1985,** *36,* 39–48.

Bomblies, K.; Wang, R.-L.; Ambrose, B. A.; Schmidt, R. J.; Meeley, R. B.; Doebley, J. Duplicate FLORICAULA/LEAFY Homologs zfl1 and zfl2 Control Inflorescence Architecture and Flower Patterning in Maize. *Development* **2003,** *130,* 2385–2395.

Camussi, A.; Jellum, M. D.; Ottaviano, E. Numerical Taxonomy of Italian Maize Populations: Fatty Acid Composition and Morphological Traits. *Maydica* **1980,** *25,* 149–165.

Gallavotti, A.; Zhao, Q.; Kyozuka, J.; Meeley, R. B.; Ritter, M. K.; Doebley, J. F.; Pè, M. E.; Schmidt, R. J. The Role of *Barren Stalk1* in the Architecture of Maize. *Nature* **2004,** *432,* 630–635.

Gallavotti, A.; Barazesh, S.; Malcomber, S.; Hall, D.; Jackson, D.; Schmidt, R. J.; McSteen, P.. *Sparse Inflorescence1* Encodes a Monocot-Specific *YUCCA*-Like Gene Required for Vegetative and Reproductive Development in Maize. *PNAS* **2008,** *105,* 15196–15201.

Garcia-Aguilar, M.; Michaud, C.; Leblanc, O.; Grimanelli, D. Inactivation of A DNA Methylation Pathway in Maize Reproductive Organs Results in Apomixis-like Phenotypes. *Plant Cell* **2010,** *22,* 3249–3267.

Gore, M. A.; Chia, J. M.; Elshire, R. J.; Sun, Q.; Ersoz, E. S.; Hurwitz, B. L.; Peiffer, J. A.; McMullen, M. D.; Grills, G. S.; Ross-Ibarra, J.; Ware, D. H.; Buckler, E. S. A First-Generation Haplotype Map of Maize. *Science* **2009**, *326*, 1115–1117.

Gutiérrez, L.; Franco, J.; Crossa, J.; Abadie, T. Comparing a Preliminary Racial Classification with a Numerical Classification of the Maize Landraces of Uruguay. *Crop Sci.* **2003**, *43*, 718–723.

Hoque, M. I.; Uddin, M. N.; Fakir, M. S.; Rasel, M. Drought and Salinity affect Leaf and Root Anatomical Structures in Three Maize Genotypes. *J. Bangladesh Agril. Univ.* **2018**, *16*(1), 47–55.

Hart, J. P.; Matson, R. G.; Thompson, R. G.; Blake, M. Teosinte Inflorescence Phytolith Assemblages Mirror Zea Taxonomy. *PLoS ONE* **2011**, *6*, e18349.

Iltis, H. H.; Doebley, J. F. Taxonomy of *Zea* (Gramineae). II. Subspecific Categories in the *Zea mays* Complex and a Generic Synopsis. *Am. J. Bot.* **1980**, *67*, 994–1004.

Leiboff, S.; Li, X.; Hu, H.-C.; Todt, N.; Yang, J.; Li, X.; Yu, X.; Muehlbauer, G. J.; Timmermans, M. C. P.; Yu, J.; Schnable, P. S.; Scanlon, M. J. Genetic Control of Morphometric Diversity in the Maize Shoot Apical Meristem. *Nat. Commun.* **2015**, *6*, 8974.

Llaurado, M.; Moreno Gonzalez, J. Classification of Northern Spanish Population of Maize by Methods of Numerical Taxonomy. *Maydica* **1993**, *38*(1), 15–21.

Louette, D.; Charrier, A.; Berthaud, J. *In Situ* Conservation of Maize in Mexico Genetic Diversity and Maize Seed Management in a Traditional Community. *Econ. Bot.* **1997**, *51*, 20–38.

Mazaheri, M.; Heckwolf, M.; Vaillancourt, B.; Gage, J. L.; Burdo, B.; Heckwolf, S.; Barry, K.; Lipzen, A.; Ribeiro, C. B.; Kono, T. J. Y.; Kaeppler, H. F.; Spalding, E. P.; Hirsch, C. N.; Buell, C. R.; de Leon, N.; Kaeppler, S. M. Genome-Wide Association Analysis of Stalk Biomass and Anatomical Traits in Maize. *BMC Plant Biol.* **2019**, *19*, 45.

Ray, J. D.; Sinclair, T. R. The Effect of Pot Size on Growth and Transpiration of Maize and Soybean During Water Deficit Stress. *J. Exp. Bot.* **1998**, *49*, 1381–1386. https://doi.org/10.1093/jxb/49.325.1381.

Sekhon, R. S.; Lin, H.; Childs, K. L., Hansey, C. N.; Buell, C. R.; de Leon, N.; Kaeppler, S. M. Genome-Wide Atlas of Transcription During Maize Development. *Plant J.* **2011**, *66*, 553–556.

Sanchez, J. J.; Goodman, M. M. Relationships among the Mexican Races of Maize. *Econ. Bot.* **1992**, *46*, 72–85.

Sanchez, J. J.; Goodman, G. M. M.; Stuber, C. W. Isozymatic and Morphological Diversity in the Races of Maize of Mexico. *Econ. Bot.* **2000**, *54*, 43–59.

Singh, A.; Selvi, M. T.; Sharma, R. Sunlight-Induced Anthocyanin Pigmentation in Maize Vegetative Tissues. *J. Exp. Bot.* **1999**, *50*, 1619–1625.

Tandzi, L. N.; Bradley, G.; Mutengwa, C. Morphological Responses of Maize to Drought, Heat and Combined Stresses at Seedling Stage. *J. Biol. Sci.* **2019**, *19*, 7–16.

Vaculík, M.; Landberg, T.; Greger, M.; Luxová, M.; Stoláriková, M.; Lux, A. Silicon Modifies Root Anatomy, and Uptake and Subcellular Distribution of Cadmium in Young Maize Plants. *Ann. Bot.* **2012**, *110*, 433–443. https://doi.org/10.1093/aob/mcs039

Zhu, J.; Vos, J.; van der Werf, W.; van der Putten, P. E. L.; Evers, J. B. Early Competition Shapes Maize Whole-Plant Development in Mixed Stands. *J. Exp. Bot.* **2014**, *65*, 641–653.

CHAPTER 6

Physiological Basis of Crop Growth and Productivity

ABSTRACT

In this chapter, different growth stages of maize, namely, growth and development flowering, pollination, and grain filling are discussed. It also discusses the mineral nutrition, water relations, and other environmental impacts on the growth and productivity of maize. Moreover, it deliberates the significant research advances in the physiological basis of maize.

6.1 GROWTH AND DEVELOPMENT

Different growth stages of the maize plant are shown in Figure 6.1. The grain yields of maize in cool temperate climates are frequently less and much variable. In these climatic conditions, the low temperatures affect the maize yields as they result in the prolongation of duration of the growth period. They not only bring about a decline in the crop growth rates but also enhance the danger of frost resulting in the termination of the premature grain filling. Wilson et al. (1995) assessed the performance of a radiation- and temperature-driven maize simulation model under these climatic conditions. They had modified this model so as to assess the impacts of both temperature and solar radiation on both the maize growth and its yield, simulated under warm as well as cool climate. For improving the simulation of the model in cool climates, certain modifications were made to the simulation model. These included an altered phenology response, a decrease in radiation-use efficiency and an increase in the rate of harvest index and also a bigger time interval between silking and the initiation of grain growth during the low temperatures. This altered model was effective in being in better agreement both with the experimental independent datasets and simulated values of grain and total biomass yield in the locations of all the tropical, subtropical, and cool-temperate regions. The functionality of the model has been improved for a wider range of climates as it exhibited only 12% of the rooted mean deviations. Thus, their findings have indicated that in maize, a greater potential in yields is achievable in regions of warm climates when there is a combination of high incident radiation, low temperature, and long growth duration. These combinations might prove effective only when during the growth of maize the mean temperatures are less than 18°C but such situations are seen only in locations having cool-temperate climates.

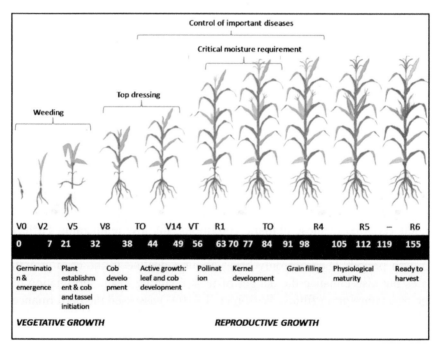

FIGURE 6.1   Growth stages of a maize plant.

6.1.1   PLASTICITY IN MAIZE DEVELOPMENT

Plasticity in the development of maize plants is chiefly under the control of some environmental signals which are perceived by the plant through some unknown signaling networks. Signaling coupled by the heterotrimeric

G-protein complex is beneath several developmental pathways. Urano et al. (2015) quantitatively assessed the morphology of two plastic developmental pathways, namely, architecture of root system and formation of female inflorescence in a mutant compact plant 2 (ct2), deficient in the alpha subunit of the heterotrimeric G-protein complex in maize. The ct2 mutant was found to compensate partially its declined shoot height by production of more leaves, more ears, and also through many pollination signals. In maize, for few plastic developmental traits, the presence of heterotrimeric G-protein complex is crucial, wherein particularly, the Gα subunit is very much essential for dampening of the excess production of female inflorescences.

6.1.2 LEAF DEVELOPMENT

Leaf anatomical heteroblastic features were identified in maize. Bongard-Pierce et al. (1996) conducted a quantitative analysis of leaf anatomy. They observed that there were parallel changes similar to that of juvenile and adult leaf traits within heteroblastic variation in cuticle thickness and epidermal cell shape in maize. Leaf blade thickness, epidermal and bundle sheath cell size, vascular area, interveinal distance, mesophyll area: bundle sheath area ratio-differed in a complex manner. They studied the genes that are involved in shoot maturation in the regulation of the above traits by examining the Teopod2 (Tp2) mutation on their expression. Tp2 increases the number of leaves that expressed the juvenile form. These included traits like cuticle thickness, epidermal cell shape, and vascular area. TP2 in other leaves led to the production of intermediate (juvenile/adult) forms with these characters. Tp2 did not have much effect on other traits. In maize, it was observed that the heteroblastic variation in the internal anatomy is mostly under the regulation of some factors which were not related to the developmental phenomenon of phase change. The Tp2 on leaf anatomy is important as it provides information about the developmental processes and their nature of development in bringing about shoot maturation. Particularly, they observed that Tp2 leaves were quantitatively intermediate between juvenile and adult leaves. It has supported the assumption that some phase-specific features of leaf identity are regulated in a combinatorial fashion rather than by mutually exclusive patterns of gene expression.

6.1.3   LEAF EXCISION AND CANOPY PHOTOSYNTHESIS

It is hypothesized that an excision of the upper strata of leaves particularly, in those plants when grown at high population densities, enhances the canopy photosynthesis apparently due to optimization of the canopy architecture and light attenuation. Liu et al. (2015) tested this hypothesis in North China plain in Jinhai 5 corn cultivar, grown at a population density of 105,000 plants ha^{-1}. Leaf excision was imposed three days after silking. When there was only the removal of uppermost and next subsequent leaf, there was an increased duration of canopy apparent photosynthesis and leaf area index (LAI), higher total dry-matter weight at physiological maturity, and even a larger harvest index value. The canopy photosynthesis duration exhibited a subsequent decline with excessive leaf removal of uppermost four and six leaves. Additionally, it was found that a modification in the number of leaves above the ear leaf resulted in varied patterns of allocation of ^{13}C photosynthates. The photosynthates were distributed to grains with excision of the top two leaves which promoted equal distribution of photosynthates to grains. However, excessive excision of leaves altered the distribution of photosynthates to grains and these were largely retained in the stem and ear bracts. Thus, those plants with only the top two leaves excised had a yield advantage at high plant density due to higher kernel weight and more harvested ears. However, the removal of four or six leaves at the very early grain-filling period (GFP), when kernel sink capacity is in the stage of establishment, resulted in a marked decline in corn yields, due to a reduction in kernel number per plant as well as kernel weight, even though there was the availability of biomass. The results have concluded that excision of two leaves in maize cultivars grown at high plant densities, delay the leaf senescence rate, and bring about an enhancement in the canopy photosynthesis and dry-matter accumulation. This would result in a larger postsilking source–sink ratio that exhibits its positive effect on the kernel weight and subsequent maize yields.

A group of undifferentiated cells are located at the shoot apical meristems controlling the initiation of the aerial plant organs in maize. There is the formation of leaves in maize all through the vegetative development. When the shoot apical meristem exhibits a transition to the floral development, tassel formation occurs. A balanced regulation occurs between that of stem cell maintenance and organogenesis. However, in the vegetative developmental stage in maize, a constraint in the development of structure and morphology of the shoot meristem occurs. The fuzzy tassel (fzt) mutant of maize possesses conspicuous inflorescence defects with indeterminate meristems, fasciation, and alterations in sex determination. These mutant plants (fzt) are short statured, with shortened plant height. They possess much narrower and shorter leaves, exhibiting leaf polarity and some defects of phase change. Thompson et al. (2014) positionally cloned fzt. The positional cloning has revealed that the fzt had a mutation in a dicer-like1 homolog, a key enzyme that was required for microRNA (miRNA) biogenesis. miRNAs are small noncoding RNAs. They bring about a reduction in the target miRNA levels. They are the chief regulators of plant development and physiology. Small RNA sequencing analysis has shown that most miRNAs are reasonably abridged in fzt plants and a small number of miRNAs are considerably condensed. They explained that the phenotype of fzt may be explained on the basis of few aspects by condensed levels of known miRNAs, together with miRNAs that control meristem determinacy, phase change, and also leaf polarity. miRNAs accountable for additional aspects of the fzt phenotype are unidentified and are probably might be those miRNAs most severely reduced in fzt mutants. The fzt mutation offers a tool to link specific miRNAs and targets to distinct phenotypes and developmental roles.

6.1.4   *NITROGEN APPLICATION ON LEAF GROWTH AND EXPANSION*

Nitrogen supply has a profound influence in maize in bringing about an increase in leaf initiation rates, its expansion, and also the leaf nitrogen contents. In tropical Australia under fully irrigated conditions, Muchow (1988) studied the effect of different nitrogen rates on the nitrogen contents of leaf, its initiation rate and expansion as well as on the leaf senescence. They observed the maize exhibited little response to the N application toward the leaf area development rather than sorghum. It had a maximum LAI of 3.8, while that of sorghum was 7.4. The differences in these leaf area developments were observed mainly because of more plant density in sorghum compared to maize. Apart from this, the vegetative development period in sorghum was much longer than that of maize. Though the variabilities in leaf initiation were moderately small, the key effect of N was observed on the leaf expansion rate. He had proposed the minimum specific leaf nitrogen concept. It is the small quantities of leaf nitrogen that is required to bring about a leaf expansion in the unit leaf area. The research results revealed that if the maize plants were able to maintain the specific leaf nitrogen above 1.0 g m^{-2} by its uptake and its allocation to leaf, then the leaf expansion rate may proceed to take place at much faster potential rates. Further, it was found that the application of nitrogen reduced the leaf senescence rate during the vegetative period. However, low rates of N application resulted in increased leaf senescence rate at the time of grain filling in the sorghum crop compared to maize. Further, it was observed that maize maintained its specific leaf nitrogen content throughout its life cycle and exhibited a decline during the grain-filling stage. Low rates of applied nitrogen increased the rate of drop in the specific leaf nitrogen content during the grain-filling stage in maize.

6.1.5   ROOT DEVELOPMENT

Root growth and development is a significant process that determines the grain yield of summer maize. When compared to the information on aboveground plant responses, inadequate information exists on the response of root growth and development to integrated agronomic practices management (defined as a complete management framework comprised of tillage method, plant density, seeding and harvest date, and fertilizer application) under field conditions. Two trials, integrated agronomic practices management (IAPM) and nitrogen rate testing (NAT) were undertaken to ascertain the effects on the summer maize root system in the course of five years in North China. IAPM comprised of four treatments (CK: local conventional cultivation practices, Opt-1: an optimized combination of cropping system and fertilizer treatment, HY: treatment based on high-yield studies, and Opt-2: a further optimized combination of cropping system and fertilizer treatment). NAT consisted of four treatments of nitrogen rate (0, 129.0, 184.5, and 300.0 kg N ha^{-1}). They measured individual/population root dry weight, individual/population absorption area, surface, volume, and length density of root and grain yield. Roots were sampled per plot at the six-leaf stage (V6), tasseling stage (VT), milk stage (R3) and physiological maturity stage (R6). The findings from IAPM showed that Opt-2 increased dry weight, volume, superficial area, and length density of root significantly through the 0–30-cm soil layer of the whole growth period. Root active absorption area of Opt-2 exhibited a significant increase in the 0–30-cm soil layer of the whole growth period except V6. NAT results showed that

N with a range from 0 to 184.5 kg N ha^{-1} performed a progressive role in root growth and development. Increment in dry weight, absorbing area of the root, and the root/shoot ratio were observed with the rising N rate within certain limits and then reduced significantly. Dry weight, the proportion of deeply distributed, absorption area, length density of root, and root/shoot ratio increased as a result of suitable population, judicious fertilizer management, and appropriate harvest date, which make available adequate nutrients and moisture to aboveground parts for growth, development, and high grain yield of summer maize (Liu et al., 2017).

Seedlings of maize (*Zea mays* L. cv WF9 × Mo17) were raised in vermiculite at different water potentials. The primary root sustained slow rates of elongation at water potentials which fully inhibited shoot growth. To get a better understanding of the root growth response, the spatial distribution of growth was examined at different water potentials. Time-lapse photography of the growth of prominent roots showed that inhibition of root elongation at low water potentials was not elucidated by a comparative reduction in growth along the length of the growing zone. Instead, longitudinal growth was unresponsive to water potentials as low as −1.6 MPa close to the root apex but was inhibited progressively in more basal locations such that the length of the growing zone reduced gradually as the water potential decreased. Cessation of longitudinal growth takes place in tissue of about the same age irrespective of spatial location or water status. Roots growing at low water potentials were also thinner, and analysis showed that radial growth rates were decreased throughout the elongation zone, resulting in significantly reduced rates of volume expansion (Sharp et al., 1988).

The maize root system includes structurally and functionally distinct root types. Mutant analyses unveiled that root-type-specific genetic regulators intrinsically control the maize root system architecture. Molecular cloning of these genes has established that major elements of auxin signal transduction, such as LOB domain (LBD) and Aux/IAA proteins, are involved in seminal, shoot-borne, and lateral root initiation. Furthermore, genetic analyses have revealed the genes associated with exocytotic vesicle docking, cell wall loosening, and cellulose synthesis and organization control root hair elongation. The detection of upstream regulators, protein interaction partners, and downstream targets of these genes with cell-type-specific transcriptome analyses have given new understandings into the regulatory networks regulating root development and architecture in maize (Hochholdinger et al., 2018).

Bengough et al. (2016) examined the physical role of root hairs in affixing the root tip during soil penetration and emergence of seedlings (Fig. 6.2). The anchorage force between the primary root of maize and the soil was measured by using a hairless maize mutant (*Z. mays*: rth 3–3) and its wild-type counterpart to find out whether root hairs allowed seedling roots in artificial biopores to penetrate sandy loam soil (dry bulk density 1.0–1.5 g cm^{-3}). Root and seedling displacements in soil adjacent to a transparent Perspex interface were analyzed by means of time-lapse imaging. Wild-type roots had five times greater (2.5 N cf. 0.5 N) peak anchorage forces than hairless mutants in 1.2 g cm^{-3} soil. Root hair anchorage allowed better soil penetration for 1.0 or 1.2 g cm^{-3} soil, but there was no significant advantage of root hairs in the densest soil (1.5 g cm^{-3}). The anchorage force was not sufficient to allow root penetration of the denser soil, may be because of less root hair penetration into pore walls and, subsequently, poorer adhesion between the root hairs and the pore walls. Hairless seedlings took 33 h to anchor themselves compared with 16 h for wild-type roots in 1.2 g cm^{-3} soil. Caryopses were frequently pushed several millimeters out of the soil before the roots became anchored and hairless roots often never became anchored securely. The physical role of root hairs in fixing the root tip may be significant in loose seedbeds above more compact soil layers and may also aid root tips to emerge from biopores and penetrate the bulk soil (Bengough et al., 2016).

FIGURE 6.2   Seedling emergence of maize (coleoptile).

Maize (*Z. mays*) forms a compound root system containing embryonic and postembryonic roots. The embryonically formed root system contains a primary root and a variable number of seminal roots. Afterward in development, the postembryonic shoot-borne root system becomes dominant and is accountable accompanied by its lateral roots for the major portion of water and nutrient uptake. Though the anatomical structure of the different root types is very similar they are initiated from different tissues during embryonic and postembryonic development. In recent times, several mutants particularly affected in maize root development have been recognized. These mutants indicated that different root-type-specific developmental programs are involved in the formation of the maize rootstock. In this review, this genetic data in the background of the maize root morphology and anatomy were summarized and a viewpoint on probable viewpoints of the molecular analysis of maize root formation was provided (Hochholdinger, 2004).

The maize (*Z. mays* L.) Aux/IAA protein RUM1 (rootless with undetectable meristems 1) regulates initiation of seminal and lateral root. To recognize RUM1-dependent gene expression patterns, RNA-Seq of the differentiation zone of primary roots of rum1 mutants and the wild type was made in four biological replicates. Overall, 2801 high-confidence maize genes exhibited differential gene expression with Fc ≥2 and FDR ≤1%. The auxin signaling-related genes rum1, like-auxin1 (lax1), lax2, (namatafcuc 1 nac1), the plethora genes plt1 (plethora 1), bbm1 (baby boom 1), and hscf1 (heat shock complementing factor 1) and the auxin response factors arf8 and arf37 were down-regulated in the mutant rum1. All of these genes except nac1 were auxin-inducible. The maize arf8 and arf37 genes are orthologs of *Arabidopsis* MP/ARF5 (MONOPTEROS/ARF5), which regulates the differentiation of vascular cells. Histological analyses of mutant rum1 roots unveiled defects in xylem organization and the variation of pith cells around the xylem. Furthermore, histochemical staining of enlarged pith cells encircling late metaxylem elements revealed that their thickened cell walls showed excessive lignin deposition. Consistent with this phenotype, rum1-dependent misexpression of several lignin biosynthesis genes was noticed. Therefore, RNA-Seq of RUM1-dependent gene expression in maize primary roots, together with histological and histochemical analyses, unveiled the specific regulation of auxin signal transduction components by RUM1 and novel functions of RUM1 in vascular development (Zhang et al., 2014).

Seminal roots of maize are important for early seedling establishment (Fig. 6.3). The rootless maize mutant relating to crown and seminal roots (rtcs) is defective in seminal root initiation during embryogenesis. Here, on the basis of histological results at three stages of seminal root primordia formation, the transcriptomes of wild-type and rtcs embryos were studied by RNA-Seq. Hierarchical clustering emphasized that samples of each genotype come together along development. Determination of their gene activity status unveiled hundreds of genes specifically transcribed in wild-type or rtcs embryos, while K-mean clustering unveiled variations in gene expression dynamics between wild type and rtcs during embryo development. Pairwise comparisons of rtcs and wild-type embryo transcriptomes recognized 131 transcription factors among 3526 differentially expressed. Among those, functional annotation emphasized genes engaged in cell-cycle control and phytohormone action, especially auxin signaling. Furthermore, in-silico promoter analyses found putative RTCS target genes related with transcription factor action and hormone metabolism and signaling. Significantly, nonsyntenic genes that appeared after the separation of maize and sorghum were overrepresented among genes exhibiting RTCS-dependent expression during seminal root primordial formation. This may put forward that these nonsyntenic genes came under the transcriptional control of the syntenic gene rtcs during seminal root evolution. On the whole, this research provided the first visions into the molecular framework fundamental to seminal root initiation in maize and makes available a primary point for further research of the molecular networks based on RTCS-dependent seminal root initiation (Tai et al., 2017).

FIGURE 6.3   Maize seedling establishment.

6.1.6 EFFECTS OF SOIL TILLAGE ON MAIZE ROOT GROWTH

It has been hypothesized that the soil tillage practices are one of the critical factors involved in the modification of the soil properties that affect root growth. The sustainability of any cropping system is dependent on the type of soil tillage practices that are adopted. An evaluation of the effects of these practices on root morphological changes and soil structural changes in a maize cropping system by Dal Ferro et al. (2014) has revealed that soil macroporosity (54–750 μm) is more affected by soil tillage practices. Conventional tillage practices resulted in a larger disruption of the structure of soil macropores and resulted in an enhanced pore class of soils possessing the soil pores of 54–250 μm, with an improvement in the loosening of the soil. A negative correlation was found to exist between the bulk density measurements and that of the indicators of root growth. Interestingly, it was found that there was only a weak effect of tillage practices on the root growth and no tillage practice has not resulted in the introduction of new architecture.

6.1.7 ROOT PHENOTYPING TECHNIQUE

In the pursuit of determining the genetic basis of root system architecture (Fig. 6.4), in recent years, advanced phenotypic techniques were developed. The phenotyping methods that were developed are precise with accurate and high outputs are also featured with control of environmental factors, which are particularly striking for mapping of the quantitative trait locus. Zurek et al. (2015) described the usage of a nondestructive *in vivo* gel-based root-imaging platform for its adaptation in exploring the root architecture of maize. They identified many contrasting traits of root system architecture from among 25 founder lines of the maize nested association mapping population. The technique could effectively locate 102 quantitative trait loci, using the B73 (compact RSA) × Ki3 (exploratory RSA) mapping population.

6.2 FLOWERING, POLLINATION, AND GRAIN FILLING

6.2.1 MAIZE FLOWERING TIME

There are many out-crossing species. Maize is one of the species which exhibits a good amount of outcrossing. In these species, adaptation to the local environments is mostly under the control of complex trait, namely, flowering time. This variation in the flowering time in maize was dissected by using 5000 recombinant inbred lines by Buckler et al. (2009). They used the maize nested association mapping population (NAM). They have assayed this variation in flowering time in approximately a million plants belonging to eight varied environments. They could not trace out any evidence of the existence of a single large-effect quantitative trait loci (QTLs). As a substitute, they evidenced and identified abundant small-effect QTLs, which were shared among the maize families. Further, they found that there was variation in the allelic effects transversely in the founder lines. They did not identify any individual QTLs, with allelic effect that was regulated by geographic origin or that with huge effects for environmental interactions or epistasis. Therefore, they mentioned that in maize flowering time can accurately be predicted by a simple additive model.

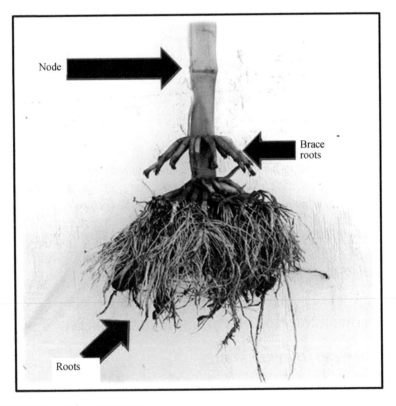

FIGURE 6.4   Maize root system.

Steinhoff et al. (2012) evaluated the genetic architecture of flowering time in an elite maize population of 684 progenies of five linked families, through joint analysis. Nine key QTLs which could explain about 50% of the genotypic variation of the flowering time trait were identified. The phenotypic disparities were much larger compared to the QTL effects on flowering time. These variations existed among the families to a large extent. Further, in a complete two-dimensional genome scan, though they did not detect any epistasis with a genetic background, four digenic–epistatic interactions were detected. Thus, their research findings have suggested that in elite maize populations, the flowering time is largely regulated by main effect QTL with pretty small effects; however, epistasis may also put into the genetic architecture of the flowering time trait.

Durand et al. (2012) in their earlier studies have found a candidate fragment length polymorphism which was seen to be in close association with the variation in flowering time in a maize inbred line F252. This fragment was detected after continuous selection of trait flowering time for a period of 7 generations in this inbred line. After its identification, in this present study, they have characterized the candidate region. The underlying polymorphisms of this candidate fragment length were identified. They have combined QTL mapping, association mapping, and developmental characterization, to dissect those genetic mechanisms which were associated with the phenotypic variation. It was found that this region in the candidate fragment had a eukaryotic initiation factor (eIF-4A). This initiation factor exposed an elevated level of sequence and structural variation beyond the 3'-UTR of eIF-4A. It included many insertions of abridged transposable elements. By the usage of a biallelic single-nucleotide polymorphism (SNP) (C/T) in the candidate region, they have established its relationship with variation in flowering time in a group of 317 inbred lines of maize. On the contrary, it was found that within the F252 genetic background the T allele was found to be associated with the late flowering time, while it was found to be associated with the early flowering time in this association panel of 317 inbred lines of maize. Three were pervasive interactions between allelic variation and the genetic background. It has pointed out the underlying epistasis. In addition, many pleiotropic effects of the candidate polymorphism were also detected on many characters, namely, flowering time, plant height, and leaf number. They broke down the relationship that existed between the flowering time and the leaf number in the progeny of a heterozygote (C/T) within the F252 background consistent with causal loci in linkage disequilibrium. Thus, they proposed that the phenotypic variation for flowering time in maize is mainly contributed by a cluster of tightly linked genes as well as epistasis.

A key trait of adaptation in maize and other crops taken as a selection criterion in the breeding of crops is the flowering time. In contrast to the earlier results of Buckler et al. (2009), Li et al. (2016) in their evaluation of flowering time variants from among 8000 maize lines of enormous large multigenetic background population in the Sino-United States environments by inclusion of two NAM panels and a natural association panel of the populations, could detect from the two-parallel linkage analysis of the 2 NAM panels, the common and also unique flowering time regions. This was found through the analyses of approximately 1 million SNPs. On the whole, maize genome they detected 90 flowering time regions. Out of these, a total of 90 flowering regions that were recognized one-third of these regions were related to the traits linked to the environmental sensitivity of the flowering time of maize. Thus, they detected 1000 flowering time associated SNPs, which were found to be distributed roughly around 220 candidate genes in the genome. These SNPs were found to be present within a distance of 1 Mb in these 220 candidate genes. Fascinatingly, they found that only two types of regions were considerably enriched for these related SNPs. Out of the two types of regions; one region was the region of candidate gene, while the other was a 5-kb region found at a distance away from that of the candidate gene. Thus, they mentioned that the flowering time was accurately predictable as most of these SNPs that were associated with the flowering time are highly accurate in nature.

Salvi et al. (2007) through the approach of positional cloning and association mapping, they found that the key flowering time quantitative trait locus, Vegetative to generative transition 1 (Vgt1) in maize is a $\approx$2-kb noncoding region, which was found to be positioned 70 kb upstream of an Ap2-like transcription factor, found to be involved in the regulation of flowering time in maize. They suggested that this Vgt1 functions as a cis-acting regulatory element. This is identified by the correlation of the Vgt1 alleles with the transcript expression levels of the downstream gene. Furthermore, they found that there are evolutionarily conserved noncoding sequences across the maize–sorghum–rice lineages within the region of Vgt1. Their findings have supported the hypotheses that variations that arise in the distant cis-acting regulatory regions are the main components for variations in plant genetic adaptation that arose during the evolution and breeding process.

6.2.2   EARLY FLOWERING IN TEMPERATE MAIZE

Maize was introduced in the southwestern United States 4000 years ago. In the lowland deserts, agriculture became established at a quick pace compared to those highlands in the temperate regions. There was a delay in the establishment of maize in agriculture in these lands for about 2000 years. Though there was a delay in the establishment of maize in temperate high lands, the temperate maize was characterized with the trait of early flowering which enabled them in faster adaptation to these regions. Swarts et al. (2017) tested this characteristic trait of early flowering of that of modern temperate maize, to identify whether the earliest upland maize was also adapted for this trait. To find the early flowering trait in the earliest maize they have sequenced 15 maize cobs which were 1900-year-old. These maize cobs were obtained from Turkey Pen Shelter in the regions of the temperate Southwest. Some genomic models have predicted that Turkey Pen maize exhibited marginal adaptation to these lands because of early flowering. They have validated these genomic models indirectly and found that Turkey Pen maize not only possessed early flowering, it was also short statured with the ability to produce tillers and exhibited high segregations for the trait of yellow-colored kernels. Thus, their findings have indicated that the differentiation found among the modern maize populations was driven by the adaptations that occurred in the temperate maize. These adaptations of temperate maize were selected *in situ* from ancient maize standing variations. Further, they have validated that if the polygenic traits are predicted, understanding of ancient phenotypes and their dynamics of adaptations to environments can be improved to a large extent.

Though teosinte was the progenitor of maize, its adaptation was confined only to the tropical environments of the regions of Mexico and that of Central America. From its center of origin, maize had a pre-Columbian spread and expanded over to the higher latitudes of America. Adaptation of maize to these regions with long-day lengths requires a post domestication selection. Several research findings have shown that there was a delay in the flowering time under long day lengths in both the tropical maize as well as in teosinte. However, it is noticed that the maize adapted to the temperate latitudes has evolved a condensed sensitivity to the length of the photoperiod. Flowering time in maize is regulated by many genes and is a complex trait influenced both by the environment and the genes. Hung et al. (2012) measured the flowering time of maize lines of nested associations and those of diverse association mapping panels under field conditions in both the photoperiods of short day length and long day lengths. They also measured the flowering time in a maize-teosinte mapping population under long day lengths. A component of flowering time in exhibiting a response toward the day lengths of photoperiod for flowering is the photoperiod response. This response involves a subset of flowering-time genes. The effects of these flowering time genes are mostly under the influence of the duration of the day length. Through the mapping of genome-wide association and targeted high-resolution linkage, they have detected a gene ZmCCT. This was found to be a homologue of the rice photoperiod response regulator Ghd7. This was found to be the key vital gene that was affecting the response of maize to flowering in accordance to the photoperiods. Their research findings have shown that in photoperiods of long day lengths, the alleles of ZmCCT obtained from varied teosintes are constantly expressed at elevated levels. The expression of these alleles conferred late flowering time than that of the alleles of maize of temperate regions. They found that most of the inbred lines of maize, even those which have been adapted to the photoperiods of the tropics, carry ZmCCT alleles that exhibit any sensitivity to day length. Thus, the indigenous farmers of the Americas were extremely successful at selecting this genetic variation of the key genes influencing the photoperiod responses and were able to develop maize varieties that have wider adaptabilities to varied environments.

6.2.3 GENES CONTROLLING FLOWERING TIME

Harnessing the diversity of landraces is not tapped completely, though most of these landraces of the domesticated species act as preservatives of the genetic variation that exists among different species. The reason behind the lag in tapping and utilization of genetic variation of these landraces is due to the presence of genetic linkages that exist between a few of the alleles that are useful and those alleles which are in large numbers and particularly are undesirable for yield improvements. Romero Navarro et al. (2017) have integrated the approaches of useful alleles and undesirable alleles and their genetic linkages and characterized the genetic diversity that existed among the 4471 landraces of maize. One thousand four hundred and ninety eight genes were identified which were found to control the latitudinal and longitudinal adaptations of these landraces.

Mapping of the genomic regions of these landraces adapted to regions of varied latitudes and longitudes led to the detection of these genes. In addition, to map the genes controlling the flowering time in maize across environments (22) usage of F-one association mapping has led to the identification of 1005 genes. Their results revealed that in totality, 61.4% of the SNPs that were found to be linked with altitude were also connected with the flowering time. Further, it was found that most of these SNPs linked to the altitude were found to be within the large structural variants in the centromeres, pericentromeric regions, and in inversions. Thus, their results of combined mapping have indicated that in maize much of the field variation in these landraces is mostly contributed due to the floral regulatory network of genes. However, more than 90% of these genes contributing to these variations are found to exhibit indirect effects.

Novel regulators of flowering time in maize were identified from a study conducted by Alter et al. (2016). They evaluated thirty inbred lines of maize to compare the time of the floral transition in these lines. The analysis of their study revealed a 3-week delay in the flowering transition in the tropical maize inbred lines compared to those inbred lines of European flint adapted to the climatic zones of temperate regions under the conditions of long days. Among these thirty inbred maize lines, four lines were found to exhibit large variations in the flowering transition. Thus, an analysis of leaf transcriptomes in these four inbred lines has shown that the flowering time trait in maize is a complex trait regulated by genes. They found that during the transition of meristem in these maize lines, the genes that are encoding the MADS box act as transcriptional regulators which are up regulated in the leaves. They demonstrated that ZmMADS1 represented the functional ortholog of the central flowering time integrator of *Arabidopsis* (SOC1). In addition, it was found that there was a delay in the flowering time in maize by RNA interference-mediated downregulation of ZmMADS1. On the contrary, a well-built overexpression of ZmMADS1 resulted in an early flowering phenotype of maize. Thus, their results indicated that ZmMADS1 has a key role in acting as a flowering time activator in maize. Thus, they reported that ZmMADS1 in maize is a positive representative of flowering time regulator in maize and does share an evolutionarily conserved function of SOC1 in *Arabidopsis.*

Maize possesses a dynamic genome. This dynamic genome does exhibit abundant transposan activity. Though it had its origin in the tropical southwestern regions of Mexico, it spread over in its cultivation over large regions of latitudinal cline in the regions of the Americas. To unveil the mystery behind the widespread latitudinal spread of maize and its adaptation, Huang et al. (2018) conducted their research studies on positional cloning and association mapping. Through these studies, they determined a flowering time quantitative trait locus similar to the Harbinger-like transposable element. This quantitative trait locus of flowering time was found to be positioned 57 kb upstream of a CCT (CONSTANS, CO-like, and TOC1) transcription factor (ZmCCT9). They found that these quantitative trait loci as Harbinger-like element acts in cis to repress ZmCCT9 expression. It promoted flowering in maize lines during the conditions of long days. To further confirm its function, they had knocked out of ZmCCT9 by CRISPR/Cas9. This caused early flowering in maize lines under conditions of long days. Additionally, they observed that this ZmCCT9 is under diurnal regulation and was found to negatively regulate the expression of the florigen ZCN8. This negative regulation of florigen ZCN8 resulted in late flowering in these lines in long days. Further, the population genetics analyses has identified that Harbinger-like transposon insertion at ZmCCT9 and the CACTA-like transposon insertion at another CCT paralog, ZmCCT10 have arisen sequentially following domestication. These two acted as targets of selection for adaptation of maize to higher latitudes. Thus, their research findings have indicated the complexity of the maize genome and its transposan activity in enabling its spread all across the latitudes.

Though maize was domesticated in southwestern Mexico, its spread and expansion over a 90° latitude in the America' required the presence of a flowering time gene enabling its adaptation across latitudes. Guo et al. (2018) has conducted a study and suggested that the ZEA CENTRORADIALIS 8 (ZCN8) gene which was identified as a key florigen gene of maize playing an important role in the mediation of flowering has a most important QTL q(DTA8). This locus that is specific for flowering time was detected constantly from numerous maize and teosinte populations analyzed. Further, from a large number of diverse inbred lines of maize, they identified a SNP (SNP-1245) in the ZCN8 promoter by following the method of association analysis. This SNP-1245 was found to be strongly linked with that of the flowering time in maize inbred lines. Additionally, this was found to be in cosegregation with qDTA8 of maize-teosinte-mapping populations. Thus, they have demonstrated that this activator is linked with the differential binding, by the flowering activator ZmMADS1 of maize. During the early domestication of maize, this SNP-1245 acted as the main target for selection and was involved in driving the preexisting allele of early flowering to near fixation in maize during its domestication.

Captivatingly, they have identified an independent association block upstream of SNP-1245. At this independent association block, the early flowering allele which was thought to be originated from *Z. mays* ssp. *mexicana* has introgressed into the early flowering haplotype of SNP-1245. This introgression enabled in the adaptation of the maize to northern high latitudes. Thus, their study demonstrated the method of selection of independent cis-regulatory variants at a gene at dissimilar evolutionary times for local adaptation. Their research findings have highlighted that cis-regulatory mechanisms are complex and mechanisms of evolution are also complex in nature. In conclusion, they have also developed and proposed a polygenic map indicating the pre-Columbian expansion of maize throughout America's high latitudes.

Thornsberry et al. (2001) used the approach of association analysis and in 92 maize inbred lines, they evaluated the Dwarf8 sequence polymorphisms. Further, by using a Bayesian analysis 4 of 141 simple sequence repeat loci, they estimated the population structure of these inbred lines. Their research findings have shown that a collection of polymorphisms are associated with the flowering time in maize and these polymorphisms are responsible for the variations that exist with reference to flowering time in maize inbreds at different regions. Further, these polymorphisms include a deletion that may change a key important domain present in that of the coding region of the gene. They further identified that there is a large distribution of nonsynonymous polymorphisms, suggesting that Dwarf8 has been a target of selection of the flowering time.

6.2.4 INDICATOR OF PHASE CHANGE IN MAIZE

In a comparative anatomy study, Sylvester et al. (2001) observed that maize juvenile leaves were coated with epicuticular wax. They did not have specialized cells, like trichomes and bulliform cells. The epidermal cell walls were uniform and stained uniformly. Similarly, the matured leaves of maize were lacking epicuticular waxes and were more pubescent. These leaves had crenulated epidermal cell walls which stained purple and blue, unlike young leaf epidermal cell walls which stained purple. Before the attainment of reproductive competence, there was an increase in leaf blade width. Similarly, there was a change in leaf blade to sheath length also. This change was continuous in the vegetative stage, which turned discontinuous much before reproduction. They found that in maize leaf primordia are initiated at a faster rate pertaining to leaf blade and sheath growth their development rate is much higher. Their research has led to the conclusion that an indicator of phase change in maize is the ratio of leaf blade to sheath, rather than leaf shape alone. Further, leaf anatomy is not the common phase change indicator.

6.2.5   FLORAL GENES IN MAIZE

Floral determinants and for the development of male and female reproductive organs, namely, tassel (Fig. 6.5) and silk (Fig. 6.6), gene AGAMOUS of *Arabidopsis* is required. Mema et al. (1996) isolated a transposon-induced mutation in ZAG1, of maize homologs of AGAMOUS, which showed a lack of determinancy. The distinctiveness of reproductive organs remained unaffected. They proposed that there is redundancy in maize sex organ specification. It has led to the detection of a second homolog *ZMM2*. Thus, these are two closely related genes having nonidentical activities.

6.2.6   REPRODUCTIVE PARTITIONING IN MAIZE

Seed number per plant can be modeled as a function of plant growth rate during the crucial period for seed set (PGR$_C$), the part of plant growth partitioned to reproductive organs (P_R) and the minimum assimilate requirement per seed (λ). Compared to PGR$_C$, very little notice is given to P_R and λ. Vega et al. (2001) analyzed reproductive partitioning and λ in three species having different reproductive strategies, namely, soybean, sunflower, and maize. Reproductive partitioning was about 50% of shoot growth in soybean and sunflower, while it was less than 50% in nonprolific maize. It was further observed that this reproductive partitioning was quite stable at different plant growths of soybean and sunflower, while it varied largely in maize. In maize, there was instability in the reproductive partitioning at plant growth rates about 2 g day^{-1}. The reproductive partitioning stability was found to be inversely associated with a PGR$_C$ threshold for reproductive growth and positively with reproductive plasticity at high PGR$_C$ among the cultivars. They further observed that maize is a species which is commonly prone to barrenness, and in this species, if the reproductive partitioning is given consideration, it enables the improvement of seed number. Seed number represented as

FIGURE 6.5 Male reproductive organ—Tassel.

FIGURE 6.6 Female reproductive organ-silk and stigma.

a function of reproductive growth is explained with hyperbolic models with x-intercepts in maize. Seed set efficiency in terms of seed number per unit of reproductive growth (Ef), increased with declining reproductive growth in maize and sunflower while it was constant in soybean. In maize, the seed set efficiency became extremely inconsistent and uneven when reproductive growth was much closer to the threshold for seed set. However, in maize, it was observed that this threshold was elevated than that of soybean or sunflower. This might be because of the presence of a greater lower combined demand for assimilate that has resulted due to a higher λ and number of concurrently developing sinks.

6.2.7 COMPUTER-BASED DETECTION OF FLOWERING STATUS OF TASSEL

The timing of tassel flowering status in maize assists in analyzing the growth status and adjustment of the farming operations accordingly, as it is one of the major cereal crops. Tassel flowering status is generally identified manually based on human observations. Lu et al. (2017) to alleviate this large-scale quantitative analysis of maize flowering status in the field environments proposed and an automatic recognition method of maize tassel flowering status. This is developed via a computer vision technology that distinguishes the tassels into nonflowering type/partially/ fully flowering type tassels (Fig. 6.7). They proposed a new metric leaning method termed large-margin dimensionality reduction. The microscopic view of pollen was seen as presented in Figure 6.8.

6.2.8 GRAIN FILLING

6.2.8.1 REMOBILIZATION OF NUTRIENTS TO GRAIN

In recent years, there was an increase in maize yields with the adoption of integrated crop soil management practices and new hybrids. Despite these increases in yield and productivity, very little attention has been given to increasing the nutrient concentrations within grains for supplementing the nutrient requirements of humans and animals as well. It is hypothesized that an increase in the retranslocation of nutrients from the vegetative organs to the grains can be effective to increase these

FIGURE 6.7 Tassel showing spikelets: sessile and pedicellate and anthers.

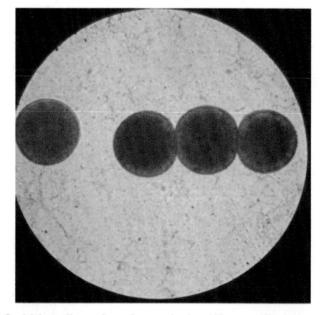

FIGURE 6.8 Maize pollen grains: microscopic view (40× magnification).

nutrient concentrations in the grains and thus the nutrient use efficiencies in crops. Chen et al. (2016) monitored dynamic changes of macro- and micronutrients in various organs of maize during the stage of grain filling. Additionally, they have also monitored the mobility of various nutrients and their contribution to maize grain nutrient content in a two-year field trial conducted at low (No N supply) and high (180 kg N ha^{-1}). It was found that under high nitrogen supply, there was high net remobilization efficiency of the vegetative organs of the nutrients N, P, K, Mn, and Zinc. The net nutrient remobilization efficiencies of these nutrients were 44%, 60%, 13%, 15%, and 25%, respectively. However, it was found that some nutrients like Mg, Ca, Fe, Cu, and B exhibited a higher rate of net accumulation in the vegetative organs as a whole, during the maize grain-filling stage. There were variations in the remobilization of these nutrients from different organs. For example, the remobilization of nutrients of N, P and Zn were from leaves with a corresponding nutrient remobilization efficiencies of 44%, 51%, and 43%, while that from stalks (including leaf sheaths and tassels) accounted for remobilization efficiencies of 48%, 71%, and 43%, respectively). Potassium was found to be remobilized to a large extent (51%) from leaves, while those of Mg, Ca, Fe, Mn, and Cu were largely remobilized from the stalks with a corresponding remobilization efficiency of 23%, 9%, 10%, 42%, and 28%, respectively. It is further observed that most of these remobilized nutrients; particularly those of Mg, Ca, Fe, Mn, Cu, and Zn were mainly translocated to the husk and cob. These two organs appeared to serve as buffer sinks for these nutrients where they accumulated. Annual climatic variations affected the remobilization of almost all nutrients except that of P, K, and Zn. Low Nitrogen stress resulted in declined remobilization efficiency of P and Zn, with an increase in Cu, while the remobilization efficiency of other nutrients remained unaltered. In the maize grain, the nutrient concentrations of P, K, Ca, Zn, and B was unchanged, or the nutrient concentrations exhibited a decline, particularly those of N, Mg, Fe, Mn, and Cu. Their research findings have concluded that the grain nutrient concentrations of N, P, K, Mn, and Zn can effectively be enhanced by the view of their increased remobilization rates from the vegetative organs, while those of Mg, Ca, Fe, Cu, and B concentration are difficult as their remobilization from vegetative organs is very less. Enhancement of leaf senescence rates under low nitrogen stress could not bring about an increase in the grain mineral nutrient concentrations. Thus, studies aiming at improving nutrient concentrations in grains by genetic improvement have to aim at increasing the nutrient remobilization, taking into consideration the organ-specific remobilization patterns of the target nutrients.

6.2.8.2   *LEAF N REMOBILIZATION DURING GRAIN FILLING*

Photosynthetic rates in maize are dependent on the amount of light harvested and rates of carbon dioxide reduction. In modern, stay green maize hybrids, it remained unclear, whether the decline in photosynthesis is due to a major contribution from the N component. Leaf nitrogen is remobilized to grains, which acts as a major contributor to grain N. However, this remobilization of N often resulted in lower photosynthetic rates. Mu et al. (2018) analyzed the relationship between remobilization of different N components and P_n during the stage of grain filling in maize at low N (no N application) and high N (180 kg N ha^{-1}). It was found that the remobilization efficiency of photosynthetic enzymes [phosphoeno/pyruvate carboxylase (PEPC), pyruvate, orthophosphate dikinase (PPDK), and Rubisco) in the leaf was several folds high than those of thylakoid N and other N components. Similarly, there was increased remobilization efficiency of all the N components of leaf under the low N supply. However, at the stage of grain filling, it was observed that there was a decline in the quantity of all the N components along with a concomitant decline in the photosynthetic rate. The ratio of P_n to the N in the PEPc, PPDK, and Rubisco appeared to increase at the grain-filling stage, while a decline in the ratio of P_n to chlorophyll and thylakoid-N occurred at this stage. Correlation analysis revealed that the photosynthetic rate was more linked to the amounts of photosynthetic enzymes rather than to chlorophyll and thylakoid N. Thus, their research findings have led to the conclusion that photosynthetic enzymes serve as an N storage reservoir at the early maize grain-filling stage. The degradation of the N storage reservoir is much crucial for the declination of the photosynthetic rate during the later stages of grain filling.

Earlier researchers have shown that the remobilization of vegetative nitrogen and that of the postsilking N both together contribute to the grain nitrogen content of maize grain. However, their regulation by sink strength of grain is less understood. Yang et al. (2016) used ^{15}N labeling for analyzing these dynamic behaviors of both pre and postsilking N, linked to source and sink manipulations in maize. It appeared that immediately after the commencement of silking, the remobilization of presilking nitrogen has begun, which appeared as a major contributor of grain nitrogen at the early stages of grain filling. However, it was found that the importance of postsilking nitrogen appeared at later stages of grain filling. It appeared that a large proportion of the postsilking nitrogen uptake was driven by the postsilking dry matter accumulated in the grain and also that of the vegetative organs. Preclusion of pollination during silking stage was found to have a minor effect on N uptake at postsilking stage. This resulted due to the compensatory growth of stems, husk + cob and roots, and leaves also continuously exported N, though there was removal of grain sink. Nitrogen requirement of grains resulted in increased remobilization efficiency of N of the leaf and that of the stem. Thus, the research findings have indicated that N remobilization from leaf is mostly under the control of sink strength rather than leaf per se. Therefore, grain N accumulation can be increased with an enhancement in the post silking N uptake, rather than the N remobilization.

6.2.8.3   GRAIN FILLING

Numerous studies conducted on the effect of heat stress on maize, have revealed that the kernel weight of the maize grains is affected largely. Edreira et al. (2014) assessed the effects of temperatures above optimal on the dynamics of biomass and water accumulation in kernels of maize hybrids with contrasting tolerance to heat stress, under field conditions. The heat stress imposed during the early phase of active grain filling had a larger effect on the kernel weight than those imposed at the silking stage or at the immediate beginning of the anthesis stage. These effects on grain filling were evident in the kernels of temperate hybrid rather than in tropical hybrid. It was observed that heat stress that was imposed at the flowering time, resulted in an enhancement of availability of photo-assimilates to each kernel at the time of this effective GFP, with a concomitant increase in the reserves of carbohydrates in the stem at the time of physiological maturity. Heat stress at flowering did not show any effect on either the dynamics of biomass or accumulation of water in kernels. A reverse trend was observed in those plants where the heat stress was imposed during the active GFP. The grain filling was interrupted to a large extent at this stage. Heat stress resulted in the establishment of strong relationships between the accumulated carbohydrate reserves in stems at the time of physiological maturity and the rate of assimilate availability per kernel during this period of effective grain filling. The effects of heat stress on kernel filling and kernel weight are more evident in the hybrid of temperate genetic background, indicating that they are extremely sensitive to heat stress.

Wilhelm et al. (1999) in their study conducted to ascertain the effects of an extended period of high temperature during the GFP on kernel growth, composition, and starch metabolism of seven inbreds of maize, revealed that there was a lengthening of the duration of grain filling on a heat unit (HU) basis under heat stress. However, an overcompensatory decline in kernel growth rate per HU resulted in an average mature kernel dry weight loss of 7% ($P = 0.06$). Similar proportional reductions were also found for starch, protein, and oil contents of the kernel. There was also a reduction in the kernel density. ADP-glucose pyrophosphorylase, glucokinase, sucrose synthase, and soluble starch synthase enzymes extracted from the endosperms were highly sensitive to high-temperature stress. Results indicated that chronic heat stress during maize grain filling reasonably restrains the seed storage processes and also the enzymes of starch metabolism.

6.2.8.4 RATE OF GRAIN FILLING

The final yields of maize are closely related to the rate of grain filling during the GFP. Jin et al. (2015) conducted a study for identifying the main miRNAs and miRNA-dependent gene regulation networks at the stage of grain filling in maize. They used a deep-sequencing technique to identify the dynamic expression patterns of miRNAs at four different developmental grain-filling stages in Zhengdan 958, an elite hybrid, cultivated largely in China. The sequencing result has shown that the expression amount of almost all miRNAs was under continuous change with the progress of the grain filling and during the development of grain. These formed into seven groups. After normalization, they codetected 77 conserved miRNAs and 74 novel miRNAs. Eighty-one out of 162 targets of the conserved miRNAs belonged to transcriptional regulation (81, 50%), followed by oxidoreductase activity (18, 11%), signal transduction (16, 10%) and development (15, 9%). The result revealed that miRNA 156, 393, 396, and 397, with their particular targets, might play important roles in the grain-filling rate (GFR) by controlling the growth, development, and environmental stress response in maize. These results have given new visions into the dynamic change of miRNAs during the developing process of the kernels of maize and assisted in the understanding of how miRNAs are functioning about the GFR.

A key trait of importance in crop domestication and of grain yields is seed size. Sekhon et al. (2014) examined the transcriptional and developmental changes that occur during seed development of maize to have an understanding of the mechanisms that are responsible in governing this trait. These changes were studied in populations divergently selected for large and small seed size from Krug, a yellow dent cultivar of maize. It was observed that, after 30 cycles of selection, seeds of the large seed population (KLS30) had a 4.7-fold higher weight and a 2.6-fold larger size than seeds of the small seed population (KSS30). The examination of patterns of accumulation of seed weight revealed that from the time of pollination through 30 days of GFP, there was an early onset, with a slower rate and earlier cessation of grain filling in KSS30 than that of KLS30. Transcriptome patterns in seeds from the populations and derived inbreds also revealed the same. In small seeds, the onset of key genes occurred at the earliest, though identical maximal transcription levels of these genes were found at the later stages even in large seeds. This has suggested that the functionally weaker alleles may be accountable to the slower rates of seed filling, to a certain extent than transcript abundance in KSS30. Further, they have identified several known genes involved in the control of cellularization and proliferation through the gene coexpression networks. In addition, many new genes were also identified which might serve as valuable candidates for biotechnological methods that are aimed at changing the seed size in maize.

6.2.9   YIELD GAPS IN MAIZE

Quantification of the changes that occur in the crop yield potentials and identification of yield gaps in a particular crop are prerequisites for determining both the yield contributing as well as yield-limiting factors to bring about enhancement in yields. Wang et al. (2014) combined simulation modeling and long-term yield records from 1981 to 2009, from 10 sites in the North China Plain to analyze the yield gaps in this region in the past three decades. APSIM simulating modeling results indicated that across these 10 sites in North China Plain the potential maize yields exhibited a declining trend even under the conditions of optimal supply of water and nitrogen. They mentioned that in these sites during the preflowering period in maize, there exists the prevalence of high temperatures coupled with declined solar radiation levels. These were the possible causes for declined yield trends. However, they mentioned that in some of these sites even though there were occurrences of high temperatures and reduced radiations the yield trends were maintained as in these sites there was continuous adoptions of new maize varieties, which assisted to maintain these preflowering periods and enabled them to be extended to postflowering periods. This coupled with an increased planting density, resulted to have a continuous increase in maize yields at these sites. Thus, there were shrinkage in maize yield gaps at all the sites tested except for the Zhengzhou site. They concluded that a future yield break through is required to overcome the yield potential gaps.

In lowland tropics, maize grain yields are reduced by short days and high temperatures. The effects of limiting environment on the genetics of maize GFR, and GFP were evaluated by Josue and Brewbaker (2018), in tropical maize germplasm, comprising of eight elite maize inbreds and 28 diallel hybrids, in three seasons at Waimanalo, Hawaii, USA. Seasonal variations comprised large variances in photosynthetic active radiation (PAR) values during the formation of maize grains. Considerable variations resulted for inbreds, hybrids, and genotype by season interactions, general combining ability (GCA) and specific combining ability (SCA) effects and their interactions with seasons. These variations were primarily attributed to the variations in the values of PAR among seasons in Hawaii at the stage of grain filling. There were predominant Additive genetic effects for GFRs and GFP. Thus, they suggested that the breeding approaches taking the benefit of additive gene effects, comprising hybrid breeding with evaluations in multiple Hawaii seasons may be used to alter GFR and GFP.

6.2.10   KERNEL GROWTH

Borrás et al. (2009) has characterized 32 and 35 inbred lines of public and elite property in 2006 and 2007, respectively, for kernel growth traits to gain the information on genotypic variability for maize (*Z. mays* L.) grain-filling patterns, at the inbred level. The results revealed a variation in kernel weight in the inbred lines across the years which exhibited a range from 104 to 317 mg kernel^{-1} and 96 to 327 mg kernel^{-1} in 2006 and 2007, respectively. Varied combinations of kernel growth rate (0.14 to 0.44 mg °Cd^{-1} kernel^{-1}) and grain-filling duration (610–1137 °Cd) resulted in the variations of kernel weight, though there was no correlation between the kernel growth rate and the grain-filling duration. A considerable genotypic variation was noticed with the moisture contents in kernels at their physiological maturity stage. The moisture concentration at this stage was in the range of 280–600 g kg^{-1}. On assessment of earlier designed framework of prediction of kernel growth on the basis of water accumulation in the kernels, it was found that genotypic variations exist for kernel water accumulation, so such a generalized framework for prediction kernel growth rate for all is not appropriate for adoption as a standard framework.

It is evident from the works of earlier researchers that in maize the kernel weight occurs in response to the changes that appear during its grain-filling stage, in terms of assimilate availability per kernel during this stage. During the grain-filling stage, plants appeared to establish and possess early kernel sink potential. This places them to grow during the late grain-filling stages in a saturating assimilate availability conditions. This implies that in maize, during the early stages of kernel development, limitations of assimilate, or source limitations are very common. Further, it is also evident that among the various maize genotypes, the reproductive efficiency in kernel setting is not a constant trait at diverse plant growth rates during the flowering stage. Gambín et al. (2006) took into consideration the plant growth rate per kernel during the GFP as a key indicator of the source availability for each kernel. Taking into consideration the plant growth rate per kernel, they tried to analyze in twelve commercial genotypes of maize, whether this plant growth rate at flowering or of the effective GFP are associated with the kernel weight and also whether any genotypic and environmental variations govern the kernel weight. A curvilinear response was found between that of kernel number per plant and plant growth rate at flowering. An increase in the plant growth rate around flowering appeared to bring about an elevated growth rate per kernel ($r^2 = 0.86$; $P < 0.001$). At around the flowering stage, considerable variations in kernel weight were found in genotypes and environments, rather than that of the effective GFP. Diverse variations were found among the genotypes in the kernel growth rates and grain-filling durations too ($P < 0.001$). The kernel growth rate appeared to exhibit its response in genotypes, with the plant growth rate per kernel which existed around flowering. The kernel growth rate did not exhibit any relationship with the plant growth rate per kernel during the effective GFP. Among the genotypes, the grain-filling duration was dependent on the ratio of the plant growth rate per kernel during the stage of effective grain filling to that of the kernel growth rate. In conclusion, the research has highlighted the importance that availability of source per kernel during the stage of early grain filling is very important and has a key role in resolving the maize kernel sink capacity and final kernel weights in cob.

Previous research works have indicated that there is a relationship between the kernel water content and duration of grain filling, wherein, the kernel water relations have a pivotal role in the control of grain-filling duration. The duration of grain filling is dependent on the relationship that exists between the water content of the kernels and the development of biomass. This relationship determines the timing of kernels reaching their critical moisture content. By the time the kernels attain critical moisture content within them, the accumulation of biomass ceases. Attainment of critical moisture content by the kernels is under the influence of the timing of cessation of kernel net water uptake or a relationship between water loss and biomass accumulation after attainment of the highest water content in the kernels. To assess the underlying mechanisms contributing to the genotypic variations in maize, Gambín et al. (2007) in 13 commercial hybrids studied the relationship that existed between that of kernel water content and volume of development, grown in diverse environments. Though these 13 commercial hybrids of maize, at the physiological maturity stage, did not exhibit any variations in their kernel percent moisture contents, they did vary in the duration of the grain filling. The duration of grain filling in these hybrids ranged from 1117 to 1470°C day. The differences in the duration of grain filling in these hybrids varied, as there were variations among these in the accumulation of thermal time units from the stage of flowering to the stage of attainment of critical percent moisture content in the kernels. These hybrids did not vary in the accumulated thermal time from the stage of silking to the attainment of maximum water content in the kernels. This stage resulted in the same kernel percent moisture content (ca. 540 g kg^{-1}). They explained the variations among these hybrids in the durations of grain filling, on the basis of the pattern of percent moisture content decline in the kernels after they have attained the maximum water content within them. It was found that there was a relationship between the percent moisture content decline or water loss and the biomass deposition in the kernels. A greater degree of water loss from the kernels resulted in speedy kernel growth, with a smaller period of grain-filling duration. The kernels attained a maximal volume after they have attained maximum water content within them, a stage nearer to the stage of physiological maturity.

Kernel development in maize exhibits a defined pattern in its development, which involves a quick increase in dry weight related to large changes in the water content within the kernels. Borrás and Westgate (2006) in their earlier researchers have identified that there was a close relationship with the maximum water content achieved by a kernel during its developmental stage of rapid grain filling with the final kernel weight. They evaluated the kernel sink capacity in the early stages of maize grain development, with reference to the percent moisture content in the kernels, in five hybrids which varied in their kernel weights. They have developed a model, which enabled in large-seeded hybrids, the determination of the relationship that existed between the dry weight accumulation as kernel weight and moisture content of kernels. A lower rate of dry-matter accumulation per unit moisture was noted in two yellow flint popcorn hybrids, which possessed small seeds. Nevertheless, a common relationship in developmental aspects of kernel growth was noted in all the genotypes evaluated, that was associated with the kernel water content and that of percent moisture content under well-watered conditions. They developed a model and coupled this developmental relationship to the final kernel weight. This model also provides a means of understanding whether any limitation of photosynthate supply at the time of kernel development has its effect on the final kernel weight.

6.2.11   *PLANT DENSITIES EFFECT ON KERNEL WEIGHT*

It is hypothesized that increased plant densities in maize have a negative effect on kernel weights due to a decline in the effective GFP. The decline in kernel weight may not be due to a decreased kernel growth rate. Previous research findings have shown that a competition arises for assimilates among the kernels, only at the later stages of grain filling. Borrás and Otegui (2001) tested this hypothesis in two commercial maize hybrids with variable kernel weights, grown at two plant densities (3 and 9 plants m^{-2}), with variable pollination treatments adopted for modification of kernel number per plant. The results revealed that there is an alteration in the kernel number per plant with a variation of the pollination treatment. Though the relationships between kernel weight and kernel number per plant were negative, such a relationship was not found to be established between years. Regression analysis revealed that kernel weight had its response to the variations in kernel number per plant. There was an increase in kernel weight (0.09–0.28 mg kernel^{-1}) per unit reduction in kernel number per plant, which varied according to plant density and genotype. Kernel weight was very much associated with variations in kernel growth rate during the effective GFP ($r^2 = 0.84$; $P < 0.001$). It was not linked to the modifications in the duration of grain filling at this stage. A source–sink ratio that was established at the time of postflowering period in each hybrid was related to the kernel weight. Nonetheless, it was also found that there were variations in the transformation capacity of these hybrids to produce biomass required for kernel weight at the postflowering period. The results thus inferred that limitations of assimilates to enable kernel growth are likely to prevail at the time of the entire GFP.

6.3   ENVIRONMENTAL IMPACTS

6.3.1   CLIMATE CHANGE IMPACT ON MAIZE YIELDS

In tropical regions, a vast wealth of historical crop data exists. As several researches have shown that climate change has its impact on yields of many crops, an understanding of its impact through new approaches is essential. Lobell et al. (2011) utilized the wealth of crop trial data that remained unexploited for climate change study. The data set comprised of more than 20,000 historical maize trials in Africa. Taking into consideration the daily weather data also combined together, they found that there was a nonlinear relationship between the maize yields and global warming. Through the analysis of the data, they found that under optimal rainfed conditions there would be a decline of 1% in the final yield with every degree day that was above 30°C, while it would be around 1.7% under conditions of drought. They specified that their results were reliable similar to those studies conducted in other regions of the world with that of temperate maize germplasm. Further, their research has indicated that the maize crop's ability to cope up with the increased heat is mostly related to the moisture content of the soil. They predicted that approximately 65% of their maize-growing areas (Africa) are likely to encounter yield losses for 1 °C of warming under optimum rain-fed management, with 100% of areas harmed by warming under drought conditions.

6.3.2 RADIATION USE EFFICIENCY IN MAIZE

In most crops plant dry matter produced particularly under nonstress environmental conditions is related directly to the quantity of the photo-synthetically active radiation that has been intercepted (IPAR) by the crop canopies. In general, in most crops, the studies that concentrate on improving this dry-matter content, commonly apply this technique in modeling. The slope of the relationship between the amount of dry matter produced and the radiation intercepted is regarded as the radiation use efficiency. This parameter appears to be constant in crop species under nonstress conditions. Kiniry et al. (1989) tested the consistency of this slope within different grain crops. The quantity of aboveground dry biomass produced per unit IPAR in maize was 3.5 g MJ^{-1} IPAR, while that in sunflower, rice, and wheat were 2.2, 2.2, and 2.8 g MJ^{-1} IPAR, respectively. Their study indicated that variabilities observed within species were not related to the variations in incident solar radiation or temperatures.

6.3.3 PRODUCTION OF HEAT SHOCK PROTEINS UNDER TEMPERATURE STRESS

Dupuis and Dumas (1990) investigated the maize reproductive tissues male and female responses to temperature stress. The fertilization capabilities of the stressed spikelets and pollen were tested by the use of *in vitro* pollination-fertilization. Through the usage of electrophoresis of ^{35}S-labeled proteins and fluorography, they analyzed the heat-shock proteins (HSPs) synthesis in these reproductive tissues and established an association among the physiological and molecular responses in maize. The investigation revealed a high rate of decline in the fertilization rate of pollinated spikelets on their exposure to a temperature more than 36°C. A separate exposure of pollen and spikelets to temperature stress has shown that the female tissues were more resistant to 4 hours of cold stress (4°C) or heat stress (40°C). There was induction of synthesis characteristic set of HSPs in the female tissues, particularly under the heat shock conditions. In contrast to the above, they noted that sensitivity to heat stress is more in the mature pollen. This sensitivity of mature pollen to heat stress is the key cause for failure of fertilization in maize at high temperatures. They could not detect any heat shock response in the matured pollen at the molecular level.

6.3.4 EFFECTS OF HIGH PLANT DENSITY

Maize is highly sensitive to plant density. For each agronomic production system, a specific plant density population is responsible for bringing about maximization of the maize yields. An overview of the factors affecting optimum plant population, with the emphasis on the influences of dense stands on ear development was presented by Sangoi (2001). The optimum plant population density of 30,000 to over 90,000 plants ha^{-1} is dependent on the factors of soil fertility, water availability, maturity, and so on, for attaining maximum economic grain yields in maize. However, the increment in the number of individuals per area beyond the optimal plant density may contribute to a series of consequences. These consequences are highly damaging to the ontogeny of the ear and often may lead to the occurrence of barrenness. It may delay the differentiation of the ear prior to its effect on the differentiation of the tassel. Apart it may also bring about a reduced growth rate in the later-initiated earshoots with only a lesser number of spikelet primordia transforming to functional florets at the flowering time. These functional florets may extrude their silks slowly. It would further result in a reduced number of fertilized spikelets because of the absence of synchronization in the anthesis and silking. Further, a restraint in carbon and nitrogen supply to the ear may bring about the stimulation in the abortion rates of young kernels after fertilization. The maize ability to face high plant populations without exhibiting excessive barrenness has been enhanced with the availability of earlier hybrids, with shorter plant height, lower leaf number, upright leaves, smaller tassels and better synchrony between male and female flowering time. Developed endurance in high stands has permitted maize to intercept and utilize solar radiation more effectively, contributing to the significant increase in grain (Sangoi, 2001).

6.4 WATER RELATIONS

A hydroponic experiment with stress of polyethylene glycol, NaCl, and 0.1 μM abscisic acid in relatively drought-tolerant cultivar of maize (*Z. mays* L. cv. Pioneer 3950) and a drought-tolerant line of sorghum (*Sorghum bicolor* L. Moench cv. ICSV 112) indicated a depression in relative growth rate by both stress factors. The relative growth rate was more depressed in maize than sorghum. In maize, it was observed that there was an increase in the growth rate by abscisic acid, which was responsible for reverting the negative effects of NaCl. The elastic modulus in maize was less than sorghum and this exhibited an increase in maize under NaCl stress. Their results indicated that variations in the elastic conditions of cell walls are responsible for limitations in growth under stress conditions (Erdei and Taleisnik, 1993).

CO_2 exchange and transpiration responses, differences in leaf water potential, and shoot and root osmolalities monitored at different stress, salt, osmolite, and drought in a maize cultivar and a drought-tolerant line of sorghum have shown that the susceptibility to these stresses was more in maize rather than sorghum. Even at low external osmolite concentration, the leaf water potential in maize showed a decline, which reached higher at higher osmolite concentrations. This decrease was in a linear manner in sorghum. Maize had a marked decline not only in its leaf water content but also in the rates of transpiration and net photosynthesis at low osmolities. These were shown to decrease gradually with an increase in osmolite concentration in sorghum. Their results indicated that the responses of these cultivars were more manifested in stomatal behavior which had its large influence on the sensitivity of CO_2 uptake to rises in external osmolities imposed either by PEG or NaCl (Nagy et al., 1995).

In Brazil, yields of maize are limited due to frequent occurrences of drought. To increase the grain production and in studying of the agronomic and physiological practices in overcoming the impact of drought stress in maize, an investigation with the role of abscisic acid in two hybrids of maize (DKB 390 and BRS 1030) resistant and sensitive has shown that there was a considerable functional relationship among RWC and the parameters of gas exchange and fluorescence during stress which were not observed by water recovery. Drought resistant cultivar DKB 390 had a higher rate of photosynthesis (P_n) and electron transport rate (ETR) than BRS 1030 at water stress and the resistant cultivar exhibited a good response to the application of ABA, which had a more content of this endogenous ABA even on the first day of water stress. This enabled the drought resistant cultivar to maintain its water status by the ABA application and minimized the effects of water stress. Further, it also enabled it to maintain an increase in the photosynthetic parameters, and a reduction in the photosystem II functions during stress (de Souza et al., 2013).

ABA catabolism has a crucial role in the ABA accumulation regulation under water deficit in maize. It accumulates in plant tissues experiencing water deficit stress. An analysis of its accumulation by its synthesis and catabolism in maize has shown that after the disappearance of the stress, when its synthesis is blocked by nordihydroguaiaretic acid, there was an 11% or more hike in the catabolic rate in comparison to that of baseline (nonstress) ABA. With such a high catabolic rate, they suggested that the xanthophyll precursor pool may likely not be able to maintain the ABA accumulation. Similarly, when fluridone was used to limit the availability of upstream ABA precursors, it was observed that ABA accumulation was sustainable only for few hours in tissues of leaf (5 h) and root (1 h). There was no such sustenance in ABA accumulation in detached tissues on the usage of fluoridone. Thus their research has suggested that the accumulation of abscisic acid within root tissues needs the import of its precursors in a regular way from shoot (Ren et al., 2007).

6.4.1   WATER DEFICIT ON LEAF APPEARANCE AND LEAF SIZE

In crop plants, the leaf growth and crop phenology are also influenced by the amount of photoperiods and the prevailing temperatures. Muchow and Carberry (1989) examined the maize tropical hybrid Dekalb XL 82 phenology and leaf growth in response to photoperiod and temperatures by the crop simulation models in the semiarid tropical environments. They obtained the mathematical relationships from the data obtained in fully irrigated grown maize crop grown at three dates of sowing. The photoperiod direct effect was found to be more on the thermal time from germination to tassel initiation. Later, it was seen on the final leaf number that developed. An exponential increase in leaf appearance was observed with thermal time. This relationship was evident in this hybrid at all times of sowings. They described the fully expanded area of every leaf as a function of leaf number. However, when taken into account the leaves that were still expanding, they observed that this association between total leaf area per plant and leaf number was identical in all the sowings, while that of leaf-area senescence and thermal time relationship was considerably variable between the sowing dates. They noted that important variations occurred when the maize crop was near to its maturity. Further, they observed that during the early vegetative growth the occurrence of water deficit has its influence on the leaf appearance rate as well as on the development of leaf size and neither on the final number of leaves developed nor on the rate of leaf senescence. At later stages of the growth of maize, the occurrence of the water deficit has its profound influence on the leaf senescence.

6.4.2 PRODUCTIVITY OF MAIZE UNDER WATER DEFICITS

Maize sorghum and pearl millet in semiarid tropics are prone to the effects of water shortages at various stages of their growth. Under these environmental conditions, the productivity of maize is mostly under the influence of its radiation use efficiency, radiation interception, and the partitioning of the dry matter to the development of the grains. Muchow (1989) made a comparative analysis of the productivities of these crops under these environments. The results indicated that maize under water shortages produced a grain yield of 6 t ha^{-1}, out yielding the other two crops, namely, sorghum, and millet under the same conditions. It failed to produce grain under water deficit conditions. However, this decline in grain production was not prominent in sorghum and millet. In response to water shortages, it was found that grain yield was more stable in millets rather than the biomass of the plant. The stability in biomass production was higher in maize and sorghum under water deficit conditions. The research studies have shown that in response to water shortages and deficits, a decline in the biomass of these crops was mostly related to a decline in the radiation use efficiency of these rather than to a decline in the interception of the radiation by the canopy of these crops except when the water stress conditions were enforced during the early vegetative growth. Maize did not exhibit much efficiency in the mobilization of its preanthesis assimilates under water deficit situations. The mobilization of these preanthesis assimilates to the development of grain was seen in sorghum and millets. Further, it was observed that under the prevalence of water deficit conditions there was conservation in the harvest index rather than the accumulation of biomass. Thus, this decline in the harvest index under conditions of water deficit had its severe impact in bringing about a decline in the grain yields.

6.4.3 MAIZE OVARY ABORTION UNDER WATER DEFICITS

Huge losses in yield arise under water deficit conditions due to larger abortions that occur in the flowers and grains. It is hypothesized that abortion in the ovary of the developing grains arises mostly due to a limitation in the carbon supply. The carbon becomes limited due to a disruption in the carbon supply that arises in the cleavage of sucrose by the invertase enzymes of cell walls in the developing ovaries. Oury et al. (2016) tested this hypothesis against another one which was linked to that of the expansion and growth of ovaries and to the silks also. In silks and ovaries of both the well-watered or moderately droughted plants, they measured the presence of abundance of those transcripts of genes which were either involved in tissue expansion rates or those concerned with the metabolism of sugars. Simultaneously, they also measured the total amount of sugars that are concentrated in these tissues along with the activities of enzymes involved in the key carbon metabolic process. During the time of water deprivation, they measured the photosynthesis and some indicators of sugar export. The results revealed that though the export of sugars is maintained by leaves, a key initial variation of molecular change was noticed in the silks prior to that in the ovaries. These molecular changes that were seen in the silks indicated that there is a variation in the genes which are involved in bringing about the expansive growth of the ovaries and silks than changes in the genes involved in the metabolism of sugars. There were large differences in the amount and concentrations of sugars that accumulated in the ovaries and also in the activities of the enzymes involved in the sugar metabolic pathways. These changes were more evident initially in the apical ovaries rather than the distal ovaries. These apical ovaries aborted at later stages. They suggested that prior to the switching of the ovaries toward abortion there was a drastic change in the concentrations of sugars and the enzyme activities. In most of the European regions, drought scenarios are very common and are characterized by moderate water deficits. Under these conditions, they proposed that these changes that are likely to occur in the carbon metabolism during the flowering period are responsible consequences for the ovary abortion rather than acting as the causes for the commencement of ovary abortion in maize. However, under severe water-deficit conditions, the ovary abortion that results might be a carbon driven one occurring at the later stages in the life cycle of maize. Thus, their research findings have also supported the view that the primary event that is responsible to bring about abortion in the reproductive organs at the end of the silking stage is the effect on the expansive growth of these organs but not an interruption in the carbon metabolism.

6.4.4   GENES AFFECTING FLOWERING TIME UNDER DROUGHT STRESS

The grain yields in the production of maize are largely dependent on the onset of flowering and drought is likely to have its impact on flowering time which is a key determinant of production. Many researches have shown that drought has its effect in causing an early arrest in the flowering of maize and leads to an increase in the anthesis-silking interval period. This increase in the anthesis-silking interval has negative effects on maize yields. Song et al. (2017) elucidated the molecular mechanisms underlying the effects of drought on ASI, by the usage of tools of RNA sequencing technology and bioinformatics. They identified 619 genes and 126 transcripts. The expression of these under short day conditions was altered to a large extent by drought stress in the leaves of B73 maize. Twenty genes were identified to exhibit their involvement in flowering time, out of the various drought responsive genes. They predicted the functions of these drought responsive genes and the transcripts by the gene ontology enrichment analysis. On the basis of analysis, they identified that the transcript levels of many genes affecting flowering time, namely, PRR37, transcription factor HY5, and CONSTANS, were considerably changed by conditions of drought. The categories analyzed under gene ontogeny which were associated with flowering time comprised the stage of reproduction, flower development, pollen–pistil interaction, and postembryonic development. In addition, they also identified numerous drought-responsive transcripts containing C_2H_2 zinc finger, three conserved cysteine residues and one histidine residue (CCCH), and NAC [the NAC acronym is derived from three genes that were initially discovered to contain a particular domain (the NAC domain): NAM (for no apical meristem), ATAF1 and −2, and CUC2 (for cup-shaped cotyledon)] domains. These domain transcripts are often involved in transcriptional regulation. Because of their involvement in transcriptional regulation, they also have immense potentiality in altering the gene expression programs involved in bringing about a change in maize flowering time. The research has provided a genome-wide analysis of differentially expressed genes, novel transcripts, and isoform variants which were expressed during the reproductive stage of maize plants when exposed to conditions of short days and drought stress.

6.4.5 INCREASE IN MAIZE OUTPUTS THROUGH IMPROVED IRRIGATION WATER SYSTEMS

Water productivity in crops is defined as the quantity of grain yield that was produced from the supply of a unit quantity of water. Water productivity benchmark is used to identify the inefficiencies of crop production and the management of water in the irrigated systems. In the Corn Belt of Western United States, out of the total maize output, 58% arises from the irrigated production. In this region, an analysis of water productivity is very meager. Grassini et al. (2011) computed the water productivity and its possibilities of increase in the maize irrigated systems in central Nebraska. On the basis of earlier studies of the associations between simulated yield and seasonal water supply, they developed a bench mark of water productivity for maize in this region. They utilized a 3-year database (2005–2007) in quantification of applied irrigation, its efficiency, and yield production. They derived a linear function from the association between simulated grain yield and seasonal water supply. The mean water productivity (WP) function (slope = 19.3 kg ha^{-1} mm^{-1}; x-intercept = 100 mm), was found to be the best benchmark for maize WP. Similarly, the average farmer's WP in central Nebraska was ~73% of the WP which has been obtained from the slope of the mean WP function. In maize, a water supply of approximately 900 mm is sufficient to bring about potential yields. However, in this region the research study had indicated that considerable (55%) of the total number of fields had excess water supply in maize production. Similarly, the pivot irrigation and conservation tillage in soybean–maize rotation fields resulted in more irrigation water-use efficiency as well as the grain yields. However, it was observed that the applied irrigation was 41% and 20% less under pivot and conservation tillage in comparison to that under surface irrigation and conventional tillage. A simulation analysis that has been carried out in the study has revealed that when the irrigation systems are shifted toward the pivot irrigation system, in this region, there could be a saving of 32% of the total yearly water volume owed to irrigated maize. Even though there might occur a little of yield decline, this could be effectively overcome with the adoption of improved synchronous irrigation schedules of crop water requirements.

6.4.6   IMPORTANCE OF N AND P FERTILIZATION AND IRRIGATION

The growth and yield traits in maize are influenced by the fertilization and irrigation schedules. The yield attributes of maize plant CV. Single-cross 10 (S.C. 10) and its grain chemical constituents were evaluated at the experimental station of NRC, Shalakan, under varying N and P fertilization rates and irrigation intervals by Ibrahim and Hala (2007). The research findings revealed increased values of plant height, ear characters, and yield of maize in both seasons at an irrigation interval of 10 days. Similarly, an increase in N–P fertilization up to 120 kg N + 35 kg P also resulted in improved growth and yield parameters. However, the interactive effect of irrigation interval and N–P fertilization could not considerably exhibit its effect on the ear weight in maize in both the seasons as well. Similar, to the enhancement of growth and yield parameters influenced by irrigation interval of 10 days and above fertilization rate, the grain chemical constituents, namely, total P mg g^{-1}, total N mg g^{-1}, crude protein %, carbohydrate %, TSS %, starch %, and oil % were also influenced in comparison to other irrigation intervals and fertilization rates and resulted in increased rates in these grain chemical constituents.

Water-use efficiency was estimated in terms of the ratio of dry-matter yield to seasonal evapotranspiration in a five-year maize irrigation field experiment. A threshold evapotranspiration of 250–300 mm or lower is responsible to bring about a decline in the production of the maize or it is quite negligible. Above this level of threshold evapotranspiration, there was a linear rise in the production rate with the application of the water. Similarly, there was also an increase in the water-use efficiency. Thus, they concluded that a "wet" irrigation regime that permits the crop to transpire at a rate that approaches the climatically induced potential with simultaneous prevention of the moisture deficits occurrence and can assist in the realization of complete and higher productivity in maize (Hillel and Guron, 1973).

6.4.7   BT PROTEINS EXPRESSION IN THE MAIZE TISSUES UNDER DROUGHT STRESS

The new transgenic maize hybrids resistant to European corn borer [*Ostrinia nubilalis* (Hübner)] commercialized by seed companies were tested for the expression of Bt proteins in the maize tissues under drought stress at the Iowa State University Hinds Irrigation Farm (Traore et al.,

2000). There was a delay in the leaf appearance up to 6 days by water deficit and reduction in plant height and leaf area by 15% and 33%, respectively. This was found to be dependent on the leaf number as well as the occurrence time of water deficit. Though many differences were not recognized for these traits in both the Bt and non-Bt maize plants, in the first generation corn borer damage, there were wide variations for these traits for the damage by second generation corn borers. The Bt plants were less affected by the damage of second-generation corn borer damage. Bt plants recorded more plant weight and grain yield (Traore et al., 2000).

6.4.8 WATER STRESS AT THE VEGETATIVE AND TASSELLING STAGES

Cakir (2004) in his three-year study of pioneer corn hybrid 3377 found that water shortages resulted in a decline in the vegetative and yield parameters. The occurrence of water stress at the vegetative and tasselling stages not only brought a decline in the plant height it also led to a drastic decline in the leaf area development. During the rapid vegetative growth period, short-duration water stress resulted in a 28–32% loss of final dry-matter weight. The omission of a single irrigation during one of the sensitive growth stages brought about a 40% loss in the grain yields, particularly during the dry years. A large extent of yield losses of 66–93% may occur if there is a prevalence of water stress prolonged during tasselling and ear formation stages. The amount of seasonal irrigation water required for non-stressed production varied by year from 390 to 575 mm.

6.5 MINERAL NUTRITION

In corn (*Z. mays* L.) seedlings affected by high P levels and low and high levels oil other nutrients, P–Zn and P–Fe associations were determined by Adriano et al. (1971). At high P and Zn levels, increased shoot growth was observed, which decreased by the deficiencies of other nutrients. N and K favored P nutrition more than Fe or Zn nutrition. Ca antagonized P, Zn, and Fe, while Mg synergized P and Zn but not Fe. In the same way, Cu and Mn antagonized shoot P and Fe concentrations and had less effect on shoot Zn concentrations. At high P levels, the most marked interaction was observed between Fe and Zn, which mutually antagonized translocation over absorption.

6.5.1 PHOSPHOROUS NUTRITION AND CARBON BUDGET

Many hypotheses were postulated pertaining to the root growth under the influence of phosphorous nutrition and the carbon budget of the plant. In maize, Mollier and Pellerin (1999) in their experimental study have found that a deprivation of P nutrient has a large and quick negative effect on the rate of leaf expansion. In the early first periods of phosphorous deficiency, there was not much reduction in the radiation use efficiency though it exhibited a reduction at the later stage. There was a reduction in the root growth, particularly; the elongation rate and the emergence of the axile roots from the first-order laterals were decreased, though there was no reduction in the density of these laterals by P starvation. These morphological responses have indicated that the availability of carbohydrates limits the formation of the roots and their growth. Their results were consistent with the assumption that P deficiency mainly affects the morphology of the root system by affecting the carbon budget of the plant with no further effect of P deficiency on root morphogenesis. The severe and early reduction of shoot growth after P deficiency may elucidate that more carbohydrates were available for root growth. This may describe the small stimulation of root growth which was detected a few days after P starvation and stated by several authors. Afterward, however, as a result of the reduced leaf area of P-deprived plants, their ability to intercept light was severely reduced so that root growth was reduced ultimately (Mollier and Pellerin, 1999).

The effect of salinity and calcium levels on water flows and on hydraulic parameters of distinct cortical cells of excised roots of young maize (*Z. mays* L. cv. Halamish) plants grown in one-third strength Hoagland solution with and without additions of NaCl and CaCl$_2$ was determined. The outcomes indicated that salinization of the growth media at regular calcium levels (0.5 mM) reduced the hydraulic conductivity Lp by three to six times, which was compensated by adding extra calcium (10 mM) to the salinized media. The reflection coefficient values (σ_s) were found near to unity, which were consistent with the statement that root cell membranes were practically not permeable to NaCl. Salinity and calcium improved the root cell diameter; the salinized seedlings, which were grown at usual calcium levels had a smaller cell length. The results demonstrated that NaCl had unfavorable effects on water transport parameters of root cells, which could be compensated with the extra additions of calcium to the medium. Similarly, their data have suggested that there was a substantial apoplasmic water flow in the root cortex and the cell-to-cell path was also responsible for the complete water transport in maize roots to bring about a reduction in the root hydraulic conductivity (Azaizeh and Steudle, 1991).

Maize hybrids varying in the prolificacy expression under conditions of different amounts of nitrogen and solar radiation that was available to each plant have shown an increase in asynchronous flowering with lessened grain yield per plant as the plant population increased from 5 to 20 m^{-1}. This decrease was found to be a resultant of a reduction of 47% in the kernel number per plant. In a semiprolific maize hybrid XL 80 and a single eared hybrid XL 81, there was a decline in the grain-yield due to nitrogen deficiency. These two hybrids exhibited 65% and 40% defoliation, respectively. Under low N conditions, they observed that there was a decline by 1.3–2.7-fold in the number of potential kernels, increased asynchronous flowering. Further, under these conditions, there was also a decline to the extent of 25% in the number of spikelets that were distinguished by the first ear. All these contributed to yield decline. Similarly, they attributed that there was an increase in the asynchronous flowering of the first ear, owing to delayed development of ears rather than the tassel and simultaneous reduction in the elongation rates of the silks in the ear. Their results have shown that their treatment effects on ear growth were more pronounced only at the lower ear positions and spikelet number was not under the direct influence of treatments (Jacobs and Pearson, 1991).

6.5.2   PHOSPHORUS UPTAKE EFFICIENCY

In maize plants, root system architecture is important not only for the acquisition of water, it has a pivotal role even in the phosphorous acquisition and other mineral nutrients. The phosphorus uptake efficiency is mostly enhanced and is dependent on the root morphological changes rather than the physiological changes that occur in the root system. A successful study case that enabled the development of P efficient maize crop on the basis of QTL selection was obtained from an investigation of the genetic relationships between the phosphorus use efficiency and root system architecture in inbred line population by Gu et al. (2016). Through the selection of some quantitative trait loci coinciding for both these traits, P-efficient lines were developed. The results have shown that there was a close correlation between the phosphorus uptake efficiency and phosphorus use efficiency and root system architecture, rather than to phosphorus utilization efficiency, particularly under low P levels. By QTL analysis, they identified a chromosome region where two QTLs for phosphorus uptake efficiency (PUE), three for phosphorus uptake efficiency, and three for root system architecture. These QTLs were assigned into two QTL clusters, Cl-bin3.04a and Cl-bin3.04b. They had positive effects from alleles derived from the large-rooted and high-phosphorus uptake efficiency parent. Further, through marker-assisted selection, they detected nine advanced backcross-derived lines which carried Cl-bin3.04a or Cl-bin3.04b. These lines in low phosphorus fields had exhibited a mean increase of 22–26% in the phosphorus use efficiency. Besides, a line L224 pyramiding Cl-bin3.04a and Cl-bin3.04b exhibited improved phosphorus uptake efficiency. This elevated phosphorus use efficiency were relied chiefly on the changes of the root morphology than the changes in root physiology.

In Brazil, savanna type of soils exhibit problems with phosphorus content. Brazilian research mainly focuses on the selection and identification of maize genotypes to such environments. In this study, the dry mass traits and characteristics of root morphology were evaluated in eight maize lines with diverse genetic background and origins. Maize lines were grown in soils with different P concentrations. The trial was conducted in plots made with two levels of phosphorus: high phosphorus and low phosphorus. The characteristics of the shoots and the root system morphology were assessed 21 days after sowing. The WinRhizo program of image analysis was used for the root morphology. The phosphorus levels did not show any differences for the dry mass attributes. But, when P levels were compared, the root morphology of the L13.1.2 strain had the highest surface area (SA) and total root length (RL), length of thin and very thin roots under low P concentration. The root systems' digital images' analysis techniques permitted the effective distinction of maize genotypes in environments with low P levels (Magalhães et al., 2011).

6.5.3 LOW NITROGEN STRESS

Nitrogen has a key role in photosynthesis and in improving crop productivity. Earlier, research findings have shown that maize plants have the ability to elevate the physiological N utilization efficiency under low-N stress by increasing photosynthetic rate (Pn) per unit leaf N, that is, photosynthetic N-use efficiency (PNUE). Mu et al. (2016) have evaluated the relationship that existed between PNUE and N allocation in the ear leaves of maize at the stage of grain filling under conditions of low nitrogen and high nitrogen conditions. The results revealed that though there was an increase in the physiological nitrogen use efficiency, there was a reduction in the grain yield under conditions of low nitrogen. Under these conditions, there was a 38% reduction in the specific leaf N content in the ear leaves, which did not have any influence on the photosynthetic rate of these ear leaves, as a result there was a 54% increase in the photosynthetic N use efficiency. It was found that under the conditions of low nitrogen stress, there was a higher investment of nitrogen into bioenergetics and sustains the electron transport. There was a decline in the allocated nitrogen to chlorophyll pigment and other light-harvesting proteins, with a corresponding decline even in the soluble proteins resulting in shrinkage of the N storage reservoir. The research findings have indicated that optimization of N allocation within leaves is one of the main adaptive mechanisms for maximization of the photosynthetic rate and crop productivity during the grain-filling stage.

6.5.4 ROOT ARCHITECTURAL AND ANATOMICAL PHENES FOR NITROGEN USE EFFICIENCY

Maize breeding in the past century had focused on yield and above ground phenes for bringing about an increase in nitrogen use efficiency, which is a vital requirement for the attainment of an increase in agricultural sustainability and food security. Drastic changes have occurred during this period in the process of maize cultivation, where in from the less fertilizer input supplies and less population densities, maize cultivation shifted over to the intensive fertilization systems and high-density populations. York et al. (2015) have hypothesized that indirect selection of the maize root traits has resulted in the evolution of phenotypes which were more suited to overcome the severe competition of nitrogen inavailability. They assessed this hypothesis in sixteen commercial maize varieties improved over the past years, by planting them at two nitrogen levels and three planting densities. The results have indicated that in the most recently developed cultivars, root systems were more or less shallow and possessed one less nodal root per whorl. These cultivars also revealed that the distance from the emergence of nodal root to that of lateral branching was double and additionally had 14% higher metaxylem vessels. However, in these cultivars, the total metaxylem vessel area has remained unaffected. This was mostly due to the prevalence of less area of 12% by every vessel of metaxylem. Plasticity has been observed even in cortical phenes, namely, aerenchyma. The aerenchyma tissue increased in the cultivars at high-density populations. Simulation modeling with SimRoot has confirmed that even the occurrence of comparatively minor variations in maize root system architecture and its anatomy could bring about an increase in the shoot growth by 16% even at high-density population or in a high nitrogen rich environment. Thus, finally they have determined that the evolution of maize root phenotypes over the last century was more consistent with ever increasing nitrogen use efficiency. Introgression of more contrasting phene states of roots into the germplasm of elite cultivars of maize, determination of the functional utility of these phene states in several agronomic conditions could add to future yield gains.

Mineral nutrients are unevenly distributed in the soil. Plants adaptability to this uneven distribution of soil nutrients determines their performance. In maize, there are different root types, namely, primary, seminal, crown, and brace roots. Yu et al. (2016) unraveled the molecular mechanisms underlying the nitrate dependent later root branching plasticity of the root types using cell-type specific transcriptome profiling and laser capture microdissection with RNA-seq at high nitrate levels. Nitrate stimulated the display of different branching patterns of these lateral roots. Among the different root types, high-level plasticity, exceptional difference from other root types under the nitrate stimulation was evident in the brace roots. Further, studies of transcriptome profiling have shown that under high nitrate simulation there is a transcriptome reprogramming in root type-specific pericycle cells, where in this alteration in the brace root pericycle cells was more significant. These have highlighted that in response to high local nitrate, maize roots exhibit differences in root-specific transcriptomes, which denotes its functional adaptability and also the systemic response to shoot nitrogen starvation in the course of development.

Similarly, to elucidate the root architectural traits for high nitrogen use efficiency, Pestsova et al. (2016) conducted a study in seven inbred maize line seedlings obtained from a southern European breeding gene pool. An adaptive response of these lines was found in all these lines for altered nitrate concentrations. A study of early plant growth under greenhouse conditions had shown that there were intergenotype differences in the nitrogen use efficiency as well as their response to water stress. These varied genotypic differences have arisen due to the association of divergent root architecture traits. Additionally, they performed mapping of quantitative trait locus for seedling root traits in 60 double haploid line populations obtained from cross NUEC2 × NUEC4 and 30 QTL for seven seedling root traits. The QTL mapping has further shown that there was clustering of few QTLs. In these two inbred lines, they found that eight loci were influencing multiple seedling root traits and were detected on chromosomes 3 and 8 overlap with the QTL for grain yield found in the same population.

6.5.5 ROOT ARCHITECTURE FOR NITROGEN ACQUISITION

In intensive cropping systems, scientists, environmental groups, and policy makers across the world are aiming at bringing about a rise in nitrogen use efficiency for the attainment of sustainable food production without any negative impacts on the environment. Nitrogen applied in high-yielding intensive maize cropping systems is lost via leaching. Mi et al. (2010) discussed the regulation of root growth by nitrogen, its movement, and uptake in the soil and root system. They mentioned that for effective acquisition of nitrate from the soil, the high yielding maize genotypes should possess the root architectural traits of a deep root system with efficient ability in extraction and uptake of nitrate from the soil much before its penetration into deeper soil layers. This is much essential particularly under large N input conditions; additionally, it should be able to put forth a dynamic lateral root growth so that it would be able to increase its effective spatial N uptake and its availability. Further, the ideotype had to exhibit a well-built response of its lateral root growth to the regions of localized nitrate supplying regions and effectively utilize this erratically distributed nitrate, under conditions of N limitations.

Maize growth and yield in low-input cropping systems are affected due to the low availability of nitrogen. Root architecture exerts an influence in the acquisition of nitrogen from soils. Maize is a crop where heterosis can be effectively utilized in the development of genotypes for effective N capture and utilization under the conditions of limited resources of Nitrogen. Chung et al. (2005) studied the relationship of root architecture and N availability in seven inbred lines and 21 of their diallel crosses. The results revealed the presence of wide genotypic variation in the root architecture. Additionally, under the less availability of Nitrogen, in these lines, there was suppression of shoot growth and increment in root to shoot ratio, with or without any concomitant increase in root biomass. Further, it was found that these exhibited their response to the low levels of nitrogen, by putting forth an increase in the total RL, with due alteration in their root architecture. The root architecture was altered with an increase in the elongation of the single axial roots and enhanced lateral root growth. Simultaneously, there was a decline in the total number of axile roots. A weak response to N deficiency in alteration of the root architecture was seen in the inbreds in production of either the root biomass or other root traits, than the crosses. Further, at both the N levels, the heterosis for root traits was significant and was accredited to both the GCA and SCA.

Under the leaching environmental conditions as most of the nitrogen losses occur through leaching, the nitrogen acquisition could be enhanced by root traits that increase the speed and efficiency of subsoil foraging. Trachsel et al. (2013) investigated the distribution of root depth in maize genotypes across the cropping cycle, effects of root angles on performance of plant, and possible plastic responses of root growth angles to nitrogen fertilization. Under high and low nitrogen field environments in the USA and South Africa, genetic variation for growth angles of crown and brace roots was studied among 108 inbred lines of maize. Rooting depth was computed as the depth containing 95% of the root mass (D95) and it showed a significant association with root angles of crown roots. Between 43 days after planting and flowering increase in the number of brace roots and rooting depth (D95) was observed. However, there were no major changes between flowering and physiological maturity.

Under well-fertilized conditions root angles of the genotypes selected as "steep" and "shallow" were not altered, while under nitrogen-deficient conditions brace and crown root angles became up to 18° steeper. Shallow genotypes showed more increases in root angles under nitrogen-deficient conditions, resulting in root angles and rooting depths comparable to the ones determined for steep genotypes. Steeper root angles allowed plastic genotypes to potentially explore similar soil volumes under nitrogen-deficient conditions as steep genotypes, thus not experiencing any decreases in grain yield compared to genotypes constitutively making steep root angles. Additive main and multiplicative interaction effect analysis showed that out of 29 genotypes best adapted to four different nitrogen fertilizer treatment-by-location combinations, 11 were steep, 11 were plastic and 7 were shallow genotypes. Among the adapted entries the number of plastic genotypes were too high compared to six that could be expected based on the distribution in the whole-genotypic set. They postulated that alteration of rooting depth by root growth angles is essential for the acquisition of nitrogen by placing roots in soil domains with the highest nitrogen availability. Genotypic variation in root growth angles and the plasticity of root growth angles in response to nitrogen may be valuable in breeding crops with better nitrogen acquisition.

The major limitation for crop production in developing nations is the suboptimal availability of nitrogen, whereas in rich nations; intensive N fertilization carries substantial environmental and economic costs. So, knowledge on root phenes that improve acquisition of N is of substantial significance. Structural-functional modeling predicted that root cortical aerenchyma (RCA) could increase the acquisition of N in maize (*Z. mays*). Here, the usefulness of RCA for N acquisition by comparing maize recombinant inbred lines with distinct RCA grown under suboptimal and adequate N availability in greenhouse mesocosms and the field at the physiological level. Under N stress conditions, the formation of RCA increased by 200% in mesocosms and by 90–100% in the field. RCA formation considerably decreased root respiration and root N content. Under low-N conditions, RCA formation increased rooting depth by 15–31%, increased leaf N content by 28–81%, increased leaf chlorophyll content by 22%, increased leaf CO_2 assimilation by 22%, increased vegetative biomass by 31–66%, and increased grain yield by 58%. The study findings were in agreement with the assumption that RCA progresses plant growth under low N conditions by reducing root metabolic costs, thus increasing soil exploration and N acquisition in deep soil strata. Though possible fitness adjustments of RCA formation are not understood well, increased RCA formation seems to be a promising breeding target for improving crop N acquisition (Saengwilai et al., 2014).

Root breeding has been put forward as an important factor in the "second green revolution" for improving crop yield and the effective use of nutrient and water resources. But, only some studies have made evident that the genetic improvement of root characteristics directly increases nutrient-use-efficiency in crops. Here, the contribution of root growth improvement to effective nitrogen acquisition and grain yield was evaluated under two different N levels. Two near-isogenic maize test crosses, T-213 (large-root) and T-Wu312 (small-root), obtained from a backcross of a BC4 F3 population from two parents (Ye478 and Wu312) with distinct root size were used for the study. T-213 had 9.6–19.5% higher RL density, root SA, and dry weight at the silking stage when compared with the control T-Wu312. There were no significant differences in the root distribution pattern in the soil profile between the two genotypes. The overall increase in root growth in T-213 improved postsilking N uptake, which increased grain yield by 17.3%. Similarly, T-213 showed a reduction in soil nitrate concentrations in the >30-cm soil layer under the high N treatment. These positive effects take place under both adequate and inadequate N supply and diverse weather conditions. This study successfully put forward that increasing root size through genetic manipulation contributes directly to effective N-uptake and higher yield (Mu et al., 2015).

6.5.6 BIOFORTIFICATION

Half of the world's population is suffering with undernourishment due to a lack of micronutrients. In making an effort to increase micronutrient levels in maize and wheat via conventional plant breeding, Ortiz-Monasterio et al. (2007) mentioned that it is vital to find genetic resources with high levels of the targeted micronutrients to consider the heritability of the targeted traits, to explore the accessibility of high throughput screening tools and to get a better understanding of genotype by environment interactions. The trait combinations in biofortified maize and wheat varieties should encourage adoption, namely, high yield potential, disease resistance, and consumer acceptability. While describing breeding approaches and targeting micronutrient levels, researchers need to consider the desired micronutrient increases, food intake, and retention and bioavailability as they relate to food processing, the presence of antinutritional factors and their promoters. Finally, ex-ante studies are required to quantify the burden of micronutrient deficiency and the potential of biofortification to attain a significant improvement in human micronutrient status in the deficient target population so as to ascertain whether a biofortification program is cost-effective.

6.5.7 IRRIGATION BY WASTE WATER

The effect of wastewater irrigation was examined on the mineral composition of corn plants in a pot experiment by Al-Jaloud et al. (1995). In corn plants, the concentration of different minerals ranged 0.67–0.89% calcium (Ca), 0.38–0.58% magnesium (Mg), 0.09–1.29% sodium (Na), 0.81–1.87% nitrogen (N), 1.81–2.27% potassium (K), 0.12–0.16% phosphorus (P), 190–257 mg kg^{-1} iron (Fe), 3.5–5.6 mg kg^{-1} copper (Cu), 37.1–44.5

mg kg⁻¹ manganese (Mn), 21.6–33.6 mg kg⁻¹ zinc (Zn), 1.40–1.84 mg kg⁻¹ molybdenum (Mo), 11.0–45.7 mg kg⁻¹ lead (Pb), and 2.5–10.8 mg kg⁻¹ nickel (Ni). The concentrations of Ca, N, K, P, Cu, and Mn in corn plants were in the deficient range excluding Mg, Fe, Zn, and Al. They mentioned that wastewater can effectively be used as a safe irrigation source.

Diatloff et al. (2008) observed that Lanthanum (La) and Cerium (Ce) decreases the growth of corn. The presence of 5·0 µM Ce reduced the dry weight of corn shoots by 32%. There was a decline in the uptake of Ca, Na, Zn, and Mn by corn with an increased in La and Ce solution. In the maize roots, the concentrations of La and Ce were high and in the shoots, they increased sharply with the increment in La or Ce concentrations in solution.

6.6   RESEARCH ADVANCES IN INCREASING CROP PRODUCTIVITY

Tollenaar and Wu (1999) analyzed the physiological basis of genetic yield improvement to make available an understanding of yield potential and to provide avenues for future yield improvement. During the last five decades yield improvement of maize (*Z. mays* L.) hybrids at the rate of ≈1.5% year⁻¹ was observed in Ontario, Canada. Comparison of short-season hybrids denoting yield improvement from the late 1950s to the late 1980s exhibited that genetic yield improvement was 2.5% per year and that most of the genetic yield improvement could be accredited to improved stress tolerance. Variances in stress tolerance among older and current hybrids have been shown for high plant population density, weed interference, low night temperatures during the GFP, low soil moisture, low soil N, and a number of herbicides. Yield improvement is the outcome of more effective capture and utilization of resources, and the enhanced efficiency in resource capture and utilization of newer hybrids is often evident under stress. The greater interception of seasonal incident radiation and higher uptake of nutrients and water resulted in improved resource capture. The improved resource capture is related with increased leaf longevity, a more active root system, and a higher ratio of assimilate supply by the leaf canopy (source) and assimilate demand by the grain (sink) during the GFP. Under optimum conditions, resource use improvements have been small because leaf photosynthesis, leaf-angle distribution of the canopy, grain chemical composition, and the proportion of dry matter distributed to the grain at maturity (i.e., harvest index) have remained almost constant. Genetic improvement of maize has been accompanied by a decrease in plant-to-plant variability. This study's findings indicated that improved stress tolerance is related with lower plant-to-plant variability and that increased plant-to-plant variability results in lower stress tolerance.

In Minnesota maize (*Z. mays* L.), yields have increased from the 2010-kg ha^{-1} yield level of the pre-1930s to the current 6290 kg ha^{-1} average. This increment in yield can be accredited to a series of technological, cultural, and management practices adopted by farmers. Cardwell (1982) attempted a study of the magnitude of the changes and the relative contributions to grain yield each practice has made over the 50-year time period. The transformation from open-pollinated to hybrid corn combined with the expected 36.5 kg ha^{-1} year^{-1} genetic gain has given 58% of the yield increase. Though it has been estimated that each kilogram of applied commercial N increases yields by 18.9 kg grain kg^{-1} for 47% of the gain in yield, the effects of less manure, and reduced N from mineralized organic matter actually produces a net N effect of 19%. Twenty-three percent of yields have been increased by the use of herbicides on 93% of the hectare to get improved weed control.

Plant densities have increased by 19,130 plants ha^{-1} contributing 47.4 kg ha^{-1}/1000 plants for a total of 21% of the gain. Improved soil drainage, fall plowing, and herbicides allowed planting 10 days earlier for an average gain of 36.4 kg ha^{-1} day^{-1} and a total of 8% of the net increase. Drilling corn instead of hill dropping has contributed 8% and fall plowing has contributed 5% to current yields. Row spacing has reduced from 107 to 90 cm for a gain of 10.2 kg ha^{-1} cm or 4% of the gain. Rotation changes had no net effect on N balance with increases in soybean [*Glycine max* (L.) Merr.] offsetting decreases in forage legume hectarage. However, there has been a negative effect of more hectares of corn following corn resulting in 3% loss from corn rootworm (*Diabrotica* spp.) and 7% loss due to the interference effect of corn following corn. Corn borer (*Ostrinia nubilalis* Hubner) became a problem in the 1940s and has reduced yield potential an average of 5%. Soil erosion has reduced yield potentials by 8% over the 50 years. Other negative and unaccounted factors reduce the potential yields by 23%. Maize yields increased continuously in places where hybrid maize has been adopted, starting in the US Corn Belt in the early 1930s. The combined effect of plant breeding and improved management practices produced this gain. Fifty percent of the increase on an average may be due to management and the rest to breeding. These two tools interact so strongly that neither of them could have produced such advancement alone. However, genetic gains may have to endure a larger share of the load in future years. Hybrid traits have undergone changes over the years. Trait variations that increased resistance to biotic and abiotic stresses (e.g., drought tolerance) are many but morphological and physiological variations that raise efficiency in growth, development, and partitioning (e.g., smaller tassels) were also recorded. Some characters have not changed over the years because breeders have intended to hold them constant (e.g., grain maturity date in US Corn Belt). In some situations, they have not changed, in spite of breeders' intention to change them (e.g., harvest index). Though breeders have constantly selected for high yield, at the same time the requirement to select for overall reliability has been a driving force in the selection of hybrids with higher stress tolerance over the years. New hybrids yield more than their predecessors in unfavorable as well as favorable growing conditions. Enhancement in the capability of the maize plant to overwhelm both large and small stress bottlenecks, instead of improvement in primary productivity has been the key driving force of improved yielding ability of new hybrids (Duvick, 2005).

6.6.1 MODELS TO STUDY CROP YIELD RESPONSE TO FERTILIZERS

Cerrato and Blackmer (1990) mentioned that decisions regarding optimum rates of fertilization directly or indirectly include fitting some type of model to yield data gathered when several rates of fertilizer are applied. Though many models are utilized in describing crop yield response to fertilizers, it is rarely elucidated why one model is chosen over others. They compared and evaluated several models (linear-plus-plateau, quadratic-plus-plateau, quadratic, exponential, and square root) normally used for describing the response of corn (*Z. mays* L.) to N fertilizer for 12 site-years of data, each having 10 rates of N applied preplanting. AH models fit the data equally well when evaluated by using the R^2 statistic. All models showed comparable maximum yields, but there were striking differences among models when predicting economic optimum rates of fertilization. Mean (across all site-years) economic optimum rates of fertilization as shown by the various models ranged from 128 to 379 kg N ha^{-1} at a common fertilizer-to-corn price ratio. Statistical analyses specified that the most commonly used model, the quadratic model, did not give an acceptable explanation of the yield responses and have a tendency to show optimal rates of fertilization that were too high. The yield responses recorded in the study were best described by the quadratic-plus-plateau model. The findings showed that, particularly amid increasing concerns about the economic and environmental effects of overfertilization, the reason for selecting one model over others deserves more attention thus it has got in the past.

Stuber et al. (1987) mentioned that characters such as grain yield are polygenetically inherited and strongly affected by environment, determination of genotypic values from phenotypic expression is not accurate and improvement strategies are often based on low heritabilities. These traits can be improved with the increased knowledge of the genetic factors involved in the expression of yield. Here, the QTLs linked with grain yield and 24 yield-associated traits were identified in two F_2 populations of maize (*Z. mays* L.) using isozyme marker loci. Further, the types and magnitudes of gene effects expressed by these QTLs were assessed. About two-thirds of the associations among 17–20 marker loci and the 25 quantitative traits were significant with a large proportion of these at $P < 0.001$.

Amounts of variation accounted by genetic factors related to individual marker loci varied from less than 1% to more than 11%. Though individual marker loci accounted for comparatively small proportions of the phenotypic variation for these yield-associated traits, variations between mean phenotypic values of the two homozygous classes at certain loci were sometimes more than 16% of the population mean. Also, diverse genomic regions contributed to yield through various subsets of the yield-related traits. Predominant types of gene action varied among loci and among the 25 quantitative traits. For plant grain yield, top ear grain weight, and ear length, the gene action was mainly dominant or over dominant. However, mainly additive gene action was involved for ear number, kernel row number, and second ear grain weight.

In the United States, an increase in commercial maize yield from about 1 Mg ha^{-1} in the 1930s to about 7 Mg ha^{-1} in the 1990s was reported. Though both the genetic and agronomic-management improvements have contributed to increase the grain yield, Tollenaar and Lee (2002) stated that the genotype × management interaction resulted in most of this improvement. The genetic improvement in maize yield is not associated with yield potential per se or with heterosis per se, but it is associated with improved stress tolerance, which is constant with the improvement in the genotype × management interaction. The prospective for future yield improvement through improved stress tolerance of maize in the

United States is large, as yield potential is about three times greater than present commercial maize yields. The mechanism used by maize breeders to improve stress tolerance is not known, but they speculated that improved stress tolerance may have resulted from the selection for yield stability. Stability analyses were applied on a number of high-yielding maize hybrids, comprising three hybrids that have been involved in some of the highest maize yields recorded in producers' fields, and studied the association between yield and yield stability. The stability analyses revealed that high-yielding maize hybrids can vary in yield stability, but findings do not support the belief that yield stability and high grain yield are mutually exclusive.

The studies on the method for successful application of subsurface drip irrigation (SDI) for corn production on the deep silt loam soils were initiated at Kansas State University of the Central Great Plains, USA, in 1989. Lamm and Trooien (2003) mentioned that in comparison with more conventional forms of irrigation the use of SDI for corn can decrease irrigation water by 35–55% in the region. Irrigation frequency was not a serious issue when SDI is utilized for corn production on the deep silt loam soils of the region. A drip line spacing of 1.5 m has been found to be most economical for corn grown in 0.76 m spaced rows. Nitrogen fertigation was a very effective management tool with SDI, which helps to maximize corn grain yield while gaining high nitrogen and water-use efficiencies. The research SDI systems have been utilized since 1989 without replacement or major degradation. SDI systems enduring 10–20 years are cost competitive for corn production with the more usual forms of irrigation in the Great Plains for certain field sizes.

6.7 CONCLUSIONS

The productivity of a crop is a result of different physiological and biochemical functions, mineral nutrition, biotic, abiotic factors, in addition to genetic and agronomic practices employed. This chapter elucidates the different physiological functions promoting maize productivity. Diverse factors affect different stages of crop growth starting from germination, seedling establishment, vegetative growth, flowering, and yield. Besides, it discusses various research advances made in increasing maize productivity.

KEYWORDS

- maize physiology
- germination
- vegetative
- flowering
- mineral nutrition
- water relations
- productivity

REFERENCES

Adriano, D. C.; Paulsen, G. M.; Murphy, L. S. Phosphorus–Iron and Phosphorus–Zinc Relationships in Corn (*Zea mays* L.) Seedlings as Affected by Mineral Nutrition. *Agron. J.* **1971**, *63*, 36–39.

Alter, P.; Bircheneder, S.; Zhou, L.-Z.; Schlüter, U.; Gahrtz, M.; Sonnewald, U.; Dresselhaus, T. Flowering Time-Regulated Genes in Maize Include the Transcription Factor ZmMADS1. *Plant Physiol.* **2016**, *172*, 389–404.

Al-Jaloud, A. A.; Hussain, G.; Saati, A. J.; Karimulla, S. Effect of Wastewater Irrigation on Mineral Com-Position of Corn and Sorghum Plants in a Pot Experiment. *J. Plant Nutr.* **1995**, *18*, 1677–1692.

Azaizeh, H.; Steudle, E. Effects of Salinity on Water Transport of Excised Maize Roots. *Plant Physiol.* **1991**, *97*, 1136–1145.

Bengough, A. G.; Loades, K.; McKenzie, B. M. Root Hairs Aid Soil Penetration by Anchoring the Root Surface to Pore Walls. *J. Exp. Bot.* **2016**, *67*(4), 1071–1078.

Bongard-Pierce, D. K.; Evans, M. M. S.; Poethig, R. S. Heteroblastic Features of Leaf Anatomy in Maize and Their Genetic Regulation. *Int. J. Plant Sci.* **1996**, *157*, 331–340.

Borrás, L.; Otegui, M. E. Maize Kernel Weight Response to Postflowering Source—Sink Ratio. *Crop Sci.* **2001**, *41*, 1816–1822.

Borrás, L.; Westgate, M. E. Predicting Maize Kernel Sink Capacity Early in Development. *Field Crops Res.* **2006**, *95*, 223–233.

Borrás, L.; Zinselmeier, C.; Senior, M. L.; Westgate, M. E.; Muszynski, M. G. Characterization of Grain-Filling Patterns in Diverse Maize Germplasm. *Crop Sci.* **2009**, *49*, 999–1009.

Buckler, E. S.; Holland, J. B.; Bradbury, P. J.; Acharya, C. B.; Brown, P. J.; Browne, C.; McMullen, M. D. The Genetic Architecture of Maize Flowering Time. *Science* **2009**, *325*, 714–718.

Cakir, R. Effect of Water Stress at Different Development Stages on Vegetative and Reproductive Growth of Corn. *Field Crops Res.* **2004**, *89*, 1–16.

Chun, L.; Mi, G.; Li, J.; Chen, F.; Zhang, F. Genetic Analysis of Maize Root Characteristics in Response to Low Nitrogen Stress. *Plant Soil* **2005**, *276*, 369–382.

Cardwell, V. B. Fifty Years of Minnesota Corn Production: Sources of Yield Increase. *Agron. J.* **1982**, *74*, 984–990.

Cerrato, M. E.; Blackmer, A. M. Comparison of Models for Describing: Corn Yield Response to Nitrogen Fertilizer. *Agron. J.* **1990**, *82*, 138–143.

Chen, Q.; Mu, X.; Chen, F.; Yuan, L.; Mi, G. Dynamic Change of Mineral Nutrient Content in Different Plant Organs During the Grain Filling Stage in Maize Grown under Contrasting Nitrogen Supply. *Eur. J. Agron.* **2016**, *80*, 137–153.

Dal Ferro, N.; Sartori, L.; Simonetti, G.; Berti, A.; Morari, F. Soil Macro- and Microstructure as Affected by Different Tillage Systems and Their Effects on Maize Root Growth. *Soil Tillage Res.* **2014**, *140*, 55–65.

de Souza, T C.; Magalhães, P. C.; de Castro, E. M.; de Albuquerque, P. E. P.; Marabesi, M. A. The Influence of ABA on Water Relation, Photosynthesis Parameters, and Chlorophyll Fluorescence under Drought Conditions in Two Maize Hybrids with Contrasting Drought Resistance. *Acta Physiol. Plant.* **2013**, *35*, 515–527.

Diatloff, E.; Smith, F. W.; Asher, C. J. Effects of Lanthanum and Cerium on the Growth and Mineral Nutrition of Corn and Mungbean. *Ann. Bot.* **2008**, *101*, 971–982.

Durand, E.; Bouchet, S.; Bertin, P.; Ressayre, A.; Jamin, P.; Charcosset, A.; Dillmann, C.; Tenaillon, M. I. Flowering Time in Maize: Linkage and Epistasis at a Major Effect Locus. *Genetics* **2012**, *190*, 1547–1562.

Dupuis, I; Dumas, C. Influence of Temperature Stress on *In Vitro* Fertilization and Heat Shock Protein Synthesis in Maize (*Zea mays* L.) Reproductive Tissues. *Plant Physiol.* **1990**, *94*, 665–670.

Duvick, D. N. The Contribution of Breeding to Yield Advances in Maize (*Zea mays* L.). *Adv. Agron.* **2005**, *86*, 83–145.

Edreira, J. I. R.; Mayer, L. I.; Otegui, M. E. Heat Stress in Temperate and Tropical Maize Hybrids: Kernel Growth, Water Relations and Assimilate Availability for Grain Filling. *Field Crops Res.* **2014**, *166*, 162–172.

Erdei, L.; Taleisnik, E. Changes in Water Relation Parameters under Osmotic and Salt Stresses in Maize and Sorghum. *Physiol. Plant.* **1993**, *89*, 381–387.

Gambín, B. L.; Borrás, L.; Otegui, M. E. Source–Sink Relations and Kernel Weight Differences in Maize Temperate Hybrids. *Field Crops Res.* **2006**, *95*, 316–326.

Gambín, B. L.; Borrás, L.; Otegui, M. E. Kernel water relations and duration of grain filling in maize temperate hybrids. *Field Crops Res.* **2007**, *101*, 1–9.

Grassini, P.; Yang, H.; Irmark, S.; Thorburn, J.; Burr, C.; Cassman, K. G. High Yield Irrigated Maize in the Western US Corn Belt. II. Irrigation Management and Crop Water Productivity. *Field Crops Res.* **2011**, *120*, 133–141.

Gu, R.; Chen, F.; Long, L.; Cai, H.; Liu, Z.; Yang, J.; Yuan, L. Enhancing Phosphorus Uptake Efficiency through QTL-Based Selection for Root System Architecture in Maize. *J. Genet. Genom.* **2016**, *43*, 663–672.

Guo, L.; Wang, X.; Zhao, M.; Huang, C.; Li, C.; Li, D.; Tian, F. Stepwise cis-Regulatory Changes in ZCN8 Contribute to Maize Flowering-Time Adaptation. *Curr. Biol.* **2018**, *28*, 3005–3015.

Hillel, D.; Guron, Y. Relation between Evapotranspiration Rate and Maize Yield. *Water Resour. Res.* **1973**, *9*, 743–748.

Hochholdinger, F.; Yu, P.; Marcon, C. Genetic Control of Root System Development in Maize. *Trends Plant Sci.* **2018**, *23*, 79–88.

Hochholdinger, F.; Woll, K.; Sauer, M.; Dembinsky, D. Genetic Dissection of Root Formation in Maize (Zea mays) Reveals Root-Type Specific Developmental Programmes. *Ann. Bot.* **2004**, *93*, 359–368.

Huang, C.; Sun, H.; Xu, D.; Chen, Q.; Liang, Y.; Wang, X.; Tian, F. ZmCCT9 Enhances Maize Adaptation to Higher Latitudes. *Proc. Natl. Acad. Sci.* **2018**, *115*, E334–E341.

Hung, H. Y.; Shannon, L. M.; Tian, F.; Bradbury, P. J.; Chen, C.; Flint-Garcia, S. A.; Holland, J. B. ZmCCT and the Genetic Basis of Day-Length Adaptation Underlying the Postdomestication Spread of Maize. *Proc. Natl. Acad. Sci.* **2012**, *109*, E1913–E1921.

Ibrahim, S. A.; Hala, K. Growth, Yield and Chemical Constituents of Corn (*Zea mays* L.) as Affected by Nitrogen and Phosphors Fertilization under Different Irrigation Intervals. *J. Appl. Sci. Res.* **2007**, *3*, 1112–1120.

Jacobs, B. C.; Pearson, C. J. Potential Yield of Maize, Determined by Rates of Growth and Development of Ears. *Field Crops Res.* **1991**, *27*, 281–298.

Jin, X.; Fu, Z.; Lv, P.; Peng, Q.; Ding, D.; Li, W.; Tang, J. Identification and Characterization of MicroRNAs During Maize Grain Filling. *PLoS One* **2015**, *10*, e0125800.

Josue, A. D. L.; Brewbaker, J. L. Diallel Analysis of Grain Filling Rate and Grain Filling Period in Tropical Maize (*Zea mays* L.). *Euphytica* **2018**, *214*, 39. https://doi.org/10.1007/s10681-017-2062-6.

Kiniry, J. R.; Jones, C. A.; O'Toole, J. C.; Blanchet, R.; Cabelguenne, M.; Spanel, D. A. Radiation Use Efficiency in Biomass Accumulation Prior to Grain Filling for Five Grain Crop Species. *Field Crops Res.* **1989**, *20*, 51–64.

Lamm, F. R.; Trooien, T. P. Subsurface Drip Irrigation for Corn Production: A Review of 10 Years of Research in Kansas. Micro-irrigation: Advances in System Design and Management. *Irrigat. Sci.* **2003**, *22*, 195–200.

Li, Y.; Li, C.; Bradbury, P. J.; Liu, X.; Lu, F.; Romay, C. M.; Wang, T. Identification of Genetic Variants Associated with Maize Flowering Time Using an Extremely Large Multi-genetic Background Population. *Plant J.* **2016**, *86*, 391–402.

Liu, T.; Gu, L.; Dong, S.; Zhang, J.; Liu, P.; Zhao, B. Optimum Leaf Removal Increases Canopy Apparent Photosynthesis, ^{13}C-Photosynthate Distribution and Grain Yield of Maize Crops Grown at High Density. *Field Crops Res.* **2015**, *170*, 32–39.

Liu, Z.; Zhu, K.; Dong, S.; Liu, P.; Zhao, B.; Zhang, J. Effects of Integrated Agronomic Practices Management on Root Growth and Development of Summer Maize. *Eur. J. Agron.* **2017**, *84*, 140–151.

Lu, H.; Cao, Z.; Xiao, Y.; Fang, Z; Zhu, Y. Towards Fine-Grained Maize Tassel Flowering Status Recognition: Dataset, Theory and Practice. *Appl. Soft Comput.* **2017**, *56*, 34–45.

Lobell, D. B.; Banziger, M.; Magorokosho, C.; Vivek, B. Nonlinear Heat Effects on African Maize as Evidenced by Historical Yield Trials. *Nat. Clim. Change* **2011**, *1*, 42–45.

Magalhães, P. C.; de Souza, T. C.; Cantão, F. R. O. Early Evaluation of Root Morphology of Maize Genotypes under Phosphorus Deficiency. *Plant, Soil Environ.* **2011**, *57*, 135–138.

Mema, M.; Ambrose, B. A.; Meeley, R. B.; Briggs, S. P.; Yanofsky, M. F.; Schmidt, R. J. Diversification of C-Function Activity in Maize Flower Development. *Science* **1996**, *274*, 1537–1540.

Mi, G.; Chen, F.; Wu, Q.; Lai, N.; Yuan, L.; Zhang, F. Ideotype Root Architecture for Efficient Nitrogen Acquisition by Maize in Intensive Cropping Systems. *Sci. China Life Sci.* **2010**, *53*, 1369–1373.

Mollier, A.; Pellerin, S. Maize Root System Growth and Development as Influenced by Phosphorus Deficiency. *J. Exp. Bot.* **1999**, *50*, 487–497.

Muchow, R. C. Effect of Nitrogen Supply on the Comparative Productivity of Maize and Sorghum in a Semi-arid Tropical Environment. I. Leaf Growth and Leaf Nitrogen. *Field Crops Res.* **1988**, *18*, 1–16.

Muchow, R. C. Comparative Productivity of Maize, Sorghum and Pearl Millet in a Semi-arid Tropical Environment. II. Effect of Water Deficits. *Field Crops Res.* **1989**, *20*, 207–219.

Muchow, R. C.; Carberry, P. S. Environmental Control of Phenology and Leaf Growth in a Tropically Adapted Maize. *Field Crops Res.* **1989**, *20*, 221–236.

Mu, X.; Chen, F.; Wu, Q.; Chen, Q.; Wang, J.; Yuan, L.; Mi, G. Genetic Improvement of Root Growth Increases Maize Yield via Enhanced Post-Silking Nitrogen Uptake. *Eur. J. Agron.* **2015**, *63*, 55–61.

Mu, X.; Chen, Q.; Chen, F.; Yuan, L.; Mi, G. Within-Leaf Nitrogen Allocation in Adaptation to Low Nitrogen Supply in Maize during Grain-Filling Stage. *Front. Plant Sci.* **2016**, *7*, 699.

Mu, X.; Chen, Q.; Chen, F.; Yuan, L.; Mi, G. Dynamic Remobilization of Leaf Nitrogen Components in Relation to Photosynthetic Rate During Grain Filling in Maize. *Plant Physiol. Biochem.* **2018**, *129*, 27–34.

Nagy, Z.; Tuba, Z.; Zsoldos, F.; Erdei, L. CO_2-Exchange and Water Relation Responses of Sorghum and Maize During Water and Salt Stress. *J. Plant Physiol.* **1995**, *145*, 539–544.

Ortiz-Monasterio, J. I.; Palacios-Rojas, N.; Meng, E.; Pixley, K.; Trethowan, R. ; Peña, R. J. Enhancing the Mineral and Vitamin Content of Wheat and Maize Through Plant Breeding. *J. Cereal Sci.* **2007**, *46*, 293–307.

Oury, V.; Caldeira, C. F.; Prodhomme, D.; Pichon, J.-P.; Gibon, Y.; Tardieu, F.; Turc, O. Is Change in Ovary Carbon Status a Cause or a Consequence of Maize Ovary Abortion in Water Deficit During Flowering? *Plant Physiol.* **2016**, *171*, 997–1008.

Pestsova, E.; Lichtblau, D.; Wever, C.; Presterl, T.; Bolduan, T.; Ouzunova, M.; Westhoff, P. QTL Mapping of Seedling Root Traits Associated with Nitrogen and Water Use Efficiency in Maize. *Euphytica* **2016**, *209*, 585–602.

Ren, H.; Gao, Z.; Chen, L.; Wei, K.; Liu, J.; Fan, Y.; Davies, W. J.; Jia, W.; Zhang, J. Dynamic Analysis of ABA Accumulation in Relation to the Rate of ABA Catabolism in Maize Tissues under Water Deficit. *J. Exp. Bot.* **2007**, *58*, 211–219.

Romero Navarro, J. A.; Willcox, M.; Burgueño, J.; Romay, C.; Swarts, K.; Trachsel, S.; Buckler, E. S. A Study of Allelic Diversity Underlying Flowering-Time Adaptation in Maize Landraces. *Nat. Genet.* **2017**, *49*, 476–480.

Salvi, S.; Sponza, G.; Morgante, M.; Tomes, D.; Niu, X.; Fengler, K. A.; Tuberosa, R. Conserved Noncoding Genomic Sequences Associated with a Flowering-Time Quantitative Trait Locus in Maize. *Proc. Natl. Acad. Sci.* **2007**, *104*, 11376–11381.

Saengwilai, P.; Nord, E. A.; Chimungu, J. G.; Brown, K. M.; Lynch, J. P. Root Cortical Aerenchyma Enhances Nitrogen Acquisition from Low-Nitrogen Soils in Maize. *Plant Physiol.* **2014**, *166*(2), 726–735.

Sangoi, L. Understanding Plant Density Effects on Maize Growth and Development: An Important Issue to Maximize Grain Yield. *Ciencia Rural* **2000**, *31*, 159–168.

Sekhon, R. S.; Hirsch, C. N.; Childs, K. L.; Breitzman, M. W.; Kell, P.; Duvick, S.; Spalding, E. P.; Buell, C. R.; de Leon, N.; Kaeppler, S. M. Phenotypic and Transcriptional Analysis of Divergently Selected Maize Populations Reveals the Role of Developmental Timing in Seed Size Determination. *Plant Physiol.* **2014,** *165*(2), 658–669.

Sharp, R. E.; Silk, W. K.; Hsiao, T. C. Growth of the Maize Primary Root at Low Water Potentials. *Plant Physiol.* **1988,** *87*(1), 50–57.

Song, K.; Kim, H. C.; Shin, S.; Kim, K.-H.; Moon, J.-C.; Kim, J. Y.; Lee, B.-M. Transcriptome Analysis of Flowering Time Genes under Drought Stress in Maize Leaves. *Front. Plant Sci.* **2017,** *8*, 267.

Steinhoff, J.; Liu, W.; Reif, J. C.; Porta, G. D.; Ranc, N.; Würschum, T. Detection of QTL for Flowering Time in Multiple Families of Elite Maize. *Theor. Appl. Genet.* **2012,** *125*, 1539–1551.

Stuber, C. W.; Edwards, M. D.; Wendel, J. F. Molecular Marker-Facilitated Investigations of Quantitative Trait Loci in Maize. II. Factors Influencing Yield and Its Component Trait. *Crop Sci.* **1987,** *27*, 639–648.

Swarts, K.; Gutaker, R. M.; Benz, B.; Blake, M.; Bukowski, R.; Holland, J.; Burbano, H. A. Genomic Estimation of Complex Traits Reveals Ancient Maize Adaptation to Temperate North America. *Science* **2017,** *357*, 512–515.

Sylvester, A. W.; Parker-Clark, V.; Murray, G. A. Leaf Shape and Anatomy as Indicators of Phase Change in the Grasses: Comparison of Maize, Rice, and Bluegrass. *Am. J. Bot.* **2001,** *88*, 2157–2167.

Tai, H.; Opitz, N.; Lithio, A.; Lu, X.; Nettleton, D.; Hochholdinger, F. Non-syntenic Genes Drive RTCS-Dependent Regulation of the Embryo Transcriptome During Formation of Seminal Root Primordia in Maize (*Zea mays* L.). *J. Exp. Bot.* **2017,** *68*(3), 403–414.

Thornsberry, J. M.; Goodman, M. M.; Doebley, J.; Kresovich, S.; Nielsen, D.; Buckler, E. S. Dwarf8 Polymorphisms Associate with Variation in Flowering Time. *Nat. Genet.* **2001,** *28*, 286–289.

Thompson, B. E.; Basham, C.; Hammond, R.; Ding, Q.; Kakrana, A.; Lee, T. F.; Simon, S. A.; Meeley, R.; Meyers, B. C.; Hake, S. The Dicer-Like1 Homolog Fuzzy Tassel Is Required for the Regulation of Meristem Determinacy in the Inflorescence and Vegetative Growth in Maize. *Plant Cell* **2014,** *26*, 4702–4717.

Trachsel, S.; Kaeppler, S. M.; Brown, K. M.; Lynch, J. P. Maize Root Growth Angles Become Steeper Under Low N Conditions. *Field Crops Res.* **2013,** *140*, 18–31.

Traore, S. B.; Carlson, R. E.; Pilcher, C. D.; Rice, M. E. Bt and Non-Bt Maize Growth and Development as Affected by Temperature and Drought Stress. *Crop Sci.* **2000,** *92*, 1027–1035.

Tollenaar, M.; Lee, E. A. Yield Potential, Yield Stability and Stress Tolerance in Maize. *Field Crops Res.* **2002,** *75*, 161–169.

Tollenaar, M.; Wu, J. Yield Improvement in Temperate Maize Is Attributable to Greater Stress Tolerance. *Crop Sci.* **1999,** *39*, 1597–1604.

Urano, D.; Jackson, D.; Jones, A. M. A G Protein Alpha Null Mutation Confers Prolificacy Potential in Maize. *J. Exp. Bot.* **2015,** *66*, 4511–4515.

Vega, C. R. C.; Andrade, F. H.; Sadras, V. O. Reproductive Partitioning and Seed Set Efficiency in Soybean, Sunflower and Maize. *Field Crops Res.* **2001,** *72*, 163–175.

Wang, J.; Wang, E.; Yin, H.; Feng, L.; Zhang, J. Declining Yield Potential and Shrinking Yield Gaps of Maize in the North China Plain. *Agric. Forest Meteorol.* **2014**, *195*, 89–101.

Wilhelm, E. P.; Mullen, R. E.; Keeling, P. L.; Singletary, G. W. Heat Stress During Grain Filling in Maize. *Crop Sci.* **1999**, *39*, 1733–1741.

Wilson, D. R.; Muchow, R. C.; Murgatroyd, C. J. Model Analysis of Temperature and Solar Radiation Limitations to Maize Potential. *Field Crops Res.* **1995**, *43*, 1–18.

Yang, L.; Guo, S.; Chen, Q.; Chen, F.; Yuan, L.; Mi, G. Use of the Stable Nitrogen Isotope to Reveal the Source-Sink Regulation of Nitrogen Uptake and Remobilization During Grain Filling Phase in Maize. *PLoS One* **2016**, *11*, e0162201.

York, L. M.; Galindo-Castaneda, T.; Schussler, J. R.; Lynch, J. P. Evolution of US Maize (*Zea mays* L.) Root Architectural and Anatomical Phenes over the Past 100 Years Corresponds to Increased Tolerance of Nitrogen Stress. *J. Exp. Bot.* **2015**, *66*, 2347–2358.

Yu, P.; Baldauf, J. A.; Lithio, A.; Marcon, C.; Nettleton, D.; Li, C.; Hochholdinger, F. Root Type-Specific Reprogramming of Maize Pericycle Transcriptomes by Local High Nitrate Results in Disparate Lateral Root Branching Patterns. *Plant Physiol.* **2016**, *170*, 1783–1798.

Zhang, Y.; Paschold, A.; Marcon, C.; Liu, S.; Tai, H.; Nestler, J.; Yeh, C. T.; Opitz, N.; Lanz, C.; Schnable, P. S.; Hochholdinger, F. The Aux/IAA Gene Rum1 Involved in Seminal and Lateral Root Formation Controls Vascular Patterning in Maize (*Zea mays* L.) Primary Roots. *J. Exp. Bot.* **2014**, *65*, 4919–4930.

Zurek, P. R.; Topp, C. N.; Benfey, P. N. Quantitative Trait Locus Mapping Reveals Regions of the Maize Genome Controlling Root System Architecture. *Plant Physiol.* **2015**, *167*, 1487–1496.

CHAPTER 7

Research Advances in Abiotic Stress Management

ABSTRACT

Abiotic stresses such as drought, high and cold temperatures, salinity, waterlogging, etc. influence the growth and productivity of maize. Many research advancements have been made in studying their effects, developed techniques, and in the identification of physiological, biochemical, and molecular mechanisms of these stress factors. The research advancements that took place are briefly reviewed in this chapter.

7.1 DROUGHT STRESS

7.1.1 QUANTITATIVE TRAIT LOCI FOR DROUGHT RESISTANCE

Drought is a complex trait. Several factors influence yield. Challenging task for the performance of improved cultivars for yield under drought is because of its complexity. Nelson et al. (2007) by using the approach of functional genomics have recognized a transcription factor. This was identified from the nuclear factor Y (NF-Y) family, *AtNF-YB1*. The mechanism of this factor though seems to be undescribed, it is found to bestow a better performance of *Arabidopsis* under drought conditions. An orthologous maize transcription factor, ZmNF-YB2, showed the same activity. Under water-deficit conditions, those transgenic maize plants which showed increased expression activity of ZmNF-YB2 were more tolerant to drought. They possessed many characters that confer drought tolerance namely, chlorophyll content, stomatal conductance, leaf temperature, reduced wilting, and maintenance of photosynthesis. These stress-related adaptations in maize may add to a speck yield improvement to maize under water stress or water-limiting conditions and environments.

In many regions where maize is cultivated, a decline in the yields was observed particularly by water deficits or drought. The most damaging effect of drought on yield may appear to be prominent when it occurs during the time of flowering. Ribaut et al (1997) conducted experiments with F_3 families, obtained from a segregating F_2 population under well-watered (WW) conditions and two other water stress systems affecting flowering [(intermediate stress (IS) and severe stress (SS)]. A number of yield components were measured on equal numbers of plants per family: grain yield (GY), ear number (ENO), kernel number (KNO), and 100-kernel weight (HKWT). Correlation analysis showed that these traits were not associated with each other.

There was a 60% decline in the maize yields by drought stress. Under WW and SS conditions yields were compared and under WW conditions the best performing families were found to be consistently more influenced by stress, and the yield declines due to SS conditions were inversely proportional to the performance under drought. Furthermore, drought-tolerance index (DTI) was found to be negatively associated with yield under WW conditions. The correlation between GY under WW and SS conditions was 0.31. So, this experiment proved that the selection for yield improvement under WW conditions only would not be very effective for improvement of yield under drought.

Quantitative trait loci (QTLs) were identified for GY, ENO, and KNO by means of composite interval mapping (CIM). Major QTLs, expressing more than 13% of the phenotypic variance were not detected for any of these characters, and there were variations in their genomic positions across water regimes. The utilization of CIM permitted the assessment of QTL-by-environment interactions (Q × E) and therefore could find "stable" QTLs across drought environments. Two such QTLs for GY, on chromosomes 1 and 10, coincided with two stable QTLs for KNO. Additionally, four genomic regions were found for both GY and the anthesis-silking interval (ASI) expression. In three of these, the allelic contributions were for short ASI and GY increase, while for that on chromosome 10 the allelic contribution for short ASI corresponded to a reduction in yield. The study findings revealed that to increase yield under drought, marker-assisted selection (MAS) by means of only the QTLs involved in the expression of yield components seems not to be the best approach, and neither does MAS using only QTLs involved in the expression of ASI. Therefore, they favor a MAS strategy that considers a combination of the "best QTLs" for various characteristics. The QTLs stable across target environments denote the largest percentage possible of the phenotypic variance, and though not influence the expression of yield directly, should be involved in the expression of characteristics strongly associated with yield, such as ASI (Ribaut et al., 1997).

In the lowland humid tropics, unpredictable drought periods are very common. These bring about significant reductions in maize yields. The effects of drought on yield reductions are more severe under situations where the drought stress prevails particularly at the flowering period. In Mexico, 85 full-sib progenies of the tropical lowland population Tuxpeno cultivated on 3 soil moisture-deficit treatments have revealed that there was considerable genotype (progeny) × soil moisture-deficit interaction for GY. Under the severe moisture deficit, yield showed a positive association with the rate of leaf and stem extension, the interval between male and female flowering and the rate of foliar senescence. These indices were utilized with GY in a selection index. To test its usefulness, experimental varieties obtained from progenies selected for yield per se under the different soil moisture-deficit treatments and a divergence (index-tolerant and index-susceptible) of performance using the selection index were cultivated under conditions similar to those of the early progeny evaluation. The maize germplasm was screened for drought tolerance (Fig. 7.1) and the drought-tolerant variety (selected for yield and favorable adaptive traits) out yielded all others by 500 kg ha^{-1} under the severe moisture-deficit but not at the loss of yield under the WW conditions. Recurrent selection was practiced for these characters under similar moisture-deficits among 250 full-sib progeny in this population for three cycles. Canopy temperature, determined by infrared thermometry before flowering, was strongly correlated ($r = -0.73^{**}$) with yield under severe moisture-deficits, and was included in the selection index after the second cycle. After three cycles of improvement with 33% progeny-selection intensity, assessment under the same soil moisture-deficits exhibited that GY increased by 1.8, 7.8, and 21.6%, or 320, 420, and 410 kg ha^{-1} cycle^{-1} under the mild, medium, and severe moisture-deficits, respectively. Under WW conditions, significant changes were observed in the drought-adaptive traits, but days-to-flowering did not show any significant change (Fischer et al., 1989).

FIGURE 7.1   Screening of maize germplasm for drought resistance.

In the United States, there was a steady increase in the maize average yields. Despite the variations in the harvestable yields, there was a large improvement in the yields even under the limitations of drought stress. Yield variability may result from the differences in the environmental stresses, nitrogen inputs, effective weed management practices, etc. The sensitivity of maize yields under water-limiting environments persists due to the prevalence of variations in the input supplies and availability of improved maize lines. Despite the drought stress bringing considerable losses in yields under variable environments, the cultivation of new commercially improved maize hybrids is enabling the production of substantial maize yields every year. Under water-limited situations, several factors contribute to better plant performance. Attempts are being done for improvement of characteristic traits in maize hybrids like better partitioning of biomass to the developing ear, faster spikelet growth, and better reproductive success. An emphasis on the faster growth rate of spikelet may bring about a decrease in the number of spikelets formed on the ear that enables overall seed set by decreasing water and carbon limitations per spikelet. To find the molecular mechanisms for drought tolerance in improved maize lines, many genomic tools are being used. Novel molecular markers and comprehensive gene expression profiling approaches offer prospects to direct the continued breeding of genotypes that gives table GY under different environmental conditions (Wesley et al., 2002).

The productivity of maize, sorghum, and pearl millet under water-deficit conditions were comparatively analyzed. There was an occurrence of water stress at different growth stages. The responses of these for water stress were evaluated for the traits like radiation interception, radiation-use efficiency, and dry-matter partitioning to grain. Muchow (1989) found that under these conditions maize yield was more than the sorghum and millet with its GY of least 6 t ha^{-1}. Under water-deficit conditions, there was no production of grain in maize, though its production of biomass was more stable rather than other crops. The reduction in biomass under water deficit may be due to its association with a higher reduction in radiation-use efficiency rather than with a reduction in radiation interception, but when the water deficit was imposed during early vegetative growth, the opposite case was observed. Pre-anthesis mobilization of assimilate to grain was noticed in sorghum and millet but not in maize. Under water deficit, harvest index was more conservative than biomass accumulation; harvest index was decreased only when water deficits severely declined the GYs.

Maize production under the semiarid conditions of the Islamic Republic of Iran during the summer requires supplemental irrigation to attain maximum yields. It is under cultivation in both the season's viz. spring and autumn. It suits their existing cropping schemes more effectively. In spite of its suitability, the yield potential of maize is variable as it is more prone to the abiotic stress effects in this region. It is well known that in this crop, the most vital stage for drought effect on yield is flowering. At the flowering stage, a single day of occurrence of water stress or drought can potentially reduce the yield of maize up to 8%. Plant populations affect most growth parameters of maize under optimal growth conditions also. Bahadori et al. (2015) in their field studies found that water stress and density had a significant effect on all characteristics.

Earlier work on maize (*Zea mays* L.) primary root growth under water stress revealed that cell elongation is sustained in the apical region of the growth zone but gradually reduced further from the apex. These responses include spatially differential and coordinated regulation of osmotic adjustment, modification of cell wall extensibility, and other cellular growth processes that are essential for root growth under water stressed conditions. As the plasma membrane is the interface between the cytoplasm and the apoplast, it possibly plays key role in these responses. In this study, the developmental allocation of plasma membrane proteins was examined in the growth zone of WW and water stressed maize primary roots by means of a simplified method for enrichment of plasma membrane proteins. They found 432 proteins with differential abundances in WW and water stressed roots. Most of the variations involved region-specific patterns of response, and the characteristics of the water stress-responsive proteins suggested participation in varied biological processes, that includes modification of sugar and nutrient transport, ion homeostasis, lipid metabolism, and cell wall composition. Integration of the different, region-specific plasma membrane protein abundance patterns with outcomes from earlier physiological, transcriptomic, and cell wall proteomic studies provided new visions into root growth adaptation to water stress (Voothuluru et al., 2016).

7.1.2   DROUGHT STRESS RESPONSES

In maize, some of the constitutive stress tolerance mechanisms associated were enlarged leaf longevity, with elevated abilities in uptake of water and nutrients uptake, superior assimilate supply during grain filling, and increased grain and ear set. Banziger et al. (2002) evaluated the adaptive changes connected to drought tolerance in four tropical maize populations grown at variable N supplies in Mexico. These populations were selected for tolerance to the occurrence of mid-season drought stress and these populations produced more number of ears per plant and kernel weight across N levels. In these populations, a close association was found between the numbers of ears per plant, linked to a shorter ASI of drought-tolerant cycles. ASI was reduced to a large extent even under N stress, though it was not found to be connected to either biomass or N accumulation of plants and ears during the flowering period. Though there was no change in the individual kernel N content with selection, a decline in grain N concentrations was observed. Similarly, the kernels with more weights have resulted due to a delay in the rate of leaf senescence and an increase in the supply of assimilated during the active grain filling stage. Thus, their findings have concluded that a declined rate of ear abortion and increased assimilate supply during grain filling of maize selected for tolerance to mid-season drought, also impart tolerance to N stress. It might thus contribute to better yields and yield stability.

7.1.3 ROOT TRAITS FOR DROUGHT ADAPTATION

Manavalan et al. (2011) screened 25 diverse parental lines of maize nested association mapping panel along with the common parental line, B73, for constitutive root traits (including rooting depth and root biomass) and shoot traits conferring adaptation to drought. The findings revealed that these diverse lines had considerable variation for traits of root viz. root length, root biomass, shoot length, and leaf area. The root length on average in these lines was in the range of 17.5–106 cm. It was observed that those genotypes which possessed a deeper root system also possessed a higher biomass and leaf area. Thus, they mentioned that this natural genetic variation that is seen with in these lines could be utilized and exploited in identifying those potential QTLs which are involved in the regulation of root architecture.

Similarly, Li et al. (2015) in their comparative study in 103 maize lines on root architecture at two regimes of water conditions viz. stressed and WW, found that a higher phenotypical diversity was found to exist between the total root length and the total root surface area. The total root length was positively correlated to the total root surface area and the root forks. Root variation among these lines was to a large extent contributed by the total root length and total root surface area. These two root traits are more suited traits of selection criteria at the seedling stage of maize. A large root morphological diversity was observed in the maize lines of temperate backgrounds possessing stiff stalk and non-stiff stalk than the groups of tropical or subtropical background. Among these tested maize lines, 7, 42, 45, and 9 were classified as drought sensitive, moderately sensitive, moderately drought tolerant, and highly drought tolerant, respectively. However, it was found that seven of the nine lines which were extremely drought tolerant belonged to the tropical/subtropical background. This has suggested that germplasm of tropical/subtropical background harbors valuable genetic resources for drought tolerance. The germplasm of this background can be effectively utilized in breeding programmes aimed at improving the drought tolerance and other abiotic stress tolerance of maize.

In maize, large genetic variations exist for the root cortical cell file number. It is hypothesized that genotypes with declined root cortical cell file numbers have better tolerance abilities of drought. Such genotypes exhibit the improved drought tolerance rates due to a reduction in their metabolic costs incurred towards a larger area of soil exploration by their roots. Chimungu et al. (2014) evaluated these postulates in maize genotypes at WW and water stress conditions at greenhouse mesocosms and in the field in the United States and Malawi. The root cortical cell file number in these genotypes ranged between 6 and 19, with a greater decline under the mesocosms grown genotypes. These genotypes had a decline in rates of root respiration by (57%) per unit of their root length. Additionally, under the water stress conditions these had a deeper rooting (15–60%), larger stomatal conductance (78%), large leaf CO_2 assimilation (32%), and more shoot biomass (52–139%). On the contrary, under similar conditions of water stress, in the field, genotypes with declined cortical cell file number, exhibited deeper rooting (33–40%), lighter stem water oxygen isotope enrichment ($\delta^{18}O$) signature (28%). This has indicated that these genotypes were significantly efficient in capturing the water from the deeper soil layers. The relative water content in the leaves of these genotypes was (10 and 35%) larger, with a larger shoot biomass also (35 and 70%) at flowering. The yields were also higher by several folds than those genotypes with many cell files. Thus, their findings have supported the above hypotheses. The variation in the genotypes with cortical cell file numbers can be effectively exploited as their breeding target for improving their drought tolerance.

7.1.4   SILICON IN IMPARTING TOLERANCE TO DROUGHT

Silicon is a beneficial element occurring naturally, known to be concerned with improving crop tolerances to abiotic stresses. Amin et al. (2018) evaluated the effects of silicon application in two maize hybrids p-33h25 and FH810 at two levels of soil water contents viz. 100 and 60% field capacity. There was a reduction in the agro-morphological as well as the physiological attributes. In drought-stressed plants of both these hybrids, the application of silicon resulted in an increase in morphological traits viz. plant height, leaf number, stem diameter, cob length, gain weight, and biological yield etc. Silicon application also resulted in an increase in the photosynthetic rate and a decline in transpiration rate.

7.2   TEMPERATURE STRESS

7.2.1   HIGH TEMPERATURE STRESS

At the early stages, maize seedlings are highly susceptible to heat stress and dries when exposed to high temperatures (Fig. 7.2). Several evidences reveal the vital role of heat shock proteins (HSPs) in improving survival at high temperatures. However, there is a significant variability in the synthesis of these among and within species, it is not known why this variation occurs and to what amount of variation in HSPs among organisms may be associated with differences in thermotolerance. One option is that HSPs production confers costs and natural selection has worked toward enhancing the cost-to-benefits of HSP synthesis and accumulation. But, still the costs of this production are not determined. If HSP production confers significant nitrogen (N) costs, then Heckathorn et al. (1996) discussed that plants cultivated under low-N conditions may accumulate less HSP than high-N plants. Furthermore, if HSPs were associated with thermotolerance, then the difference in HSPs induced by N (or other factors) may associate with the disparity in thermotolerance. To test these predictions, Heckathorn et al. (1996) cultivated a variety of corn (*Zea mays* L.) under various N levels and then subjected the plants to serious heat stress. They observed that high-N plants produced more amounts of mitochondrial Hsp60 and chloroplastic Hsp24 per unit protein. In these, the patterns of HSP production were associated with PSII efficiency, which was measured by F_v/F_m. Therefore, their findings indicated that N availability affects HSP production in higher plants. This in turn suggested that production of HSP might be resource-limited and that among other benefits; chloroplast HSPs (e.g., Hsp24) may have limit damage due to PSII function during heat stress (Heckathorn et al., 1996).

Waananen and Okos (1988) experimentally determined the failure properties of kernels in hybrid corn. They performed uniaxial compressive tests of core specimens at 9.7, 14, 17, 21, and 26% dry basis (DB) moisture levels, and at the temperature levels of 25, 55, 85 and 100°C. The slope of the stress–strain curve linear portion declined with an increment in kernel moisture content and temperature. Failure strain increased as the moisture content and temperature increased. However, the modulus of toughness decreased with a rise in temperature. At a constant temperature and at a moisture content of about 16%, DB the modulus of toughness reached maximum and then decreased with a further increase in moisture content.

FIGURE 7.2   Screening for heat stress tolerance-dried seedlings (heat susceptible line).

7.2.2   *SHELTER BELTS IN MODIFYING MICROCLIMATE IN CORN BELT*

Shelterbelts (field windbreaks) are tools of importance for farming in semiarid areas. However, they are not in use largely. This is mostly due to the absence or lack of sufficient information on the site specific suitability of shelter belts, their establishment, and maintenance. A group of researchers were actively involved in developing a modeling system that will assess site-specific effects, benefits, and costs for sheltered fields that produce maize or corn (*Zea maize*) and soybean (*Glycine max*) in the United States Corn Belt region. A key element of the modeling system is the utilization of the CROPGRO-Soybean and CERES-Maize models to simulate yield response to changes in microclimatic across a sheltered field. Mize et al. (2005) tested the capability of these two models to simulate yield in a sheltered field. Both models simulated yield improvements because of shelter. The soybean model was more responsive to microclimatic variations than the maize model. Long-term simulations normally exhibited a field level increment in yield due to shelter for maize and soybeans with a mean increase of 4.1 and 3.3, respectively. Modification in wind run due to shelter is more significant in rising yield than variations in temperature.

The CERES-Maize model appears to be more sensitive to variations in wind run than the CROPGRO-Soybean mode (Mize et al., 2005).

Irrigation scheduling is progressively more key practice in the management of valuable water resources in agricultural areas. Clawson and Blad (1982) assessed the possibility of scheduling irrigation in corn (*Zea mays* L.) by means of canopy temperature data taken with a hand-held infrared thermometer (IRT). A total of 283 mm of irrigation water was applied to the WW plot while the CTV plot was applied with 127 mm. By the end of the growing season, approximately all available water had been extracted from the soil in the temperature scheduled plots but the soil profile persisted near field capacity in the WW plot. This shows that during the growing season soil water was most efficiently utilized in the temperature scheduled plots. Grain yields (Y) were decreased in the order of increasing water stress in this way: $Y_{WW} > Y_{CTV} > Y_{1SMW} > Y_{3SMW} > Y_{DL}$. Y_{DL} is the yield from a dry land (DL) plot in which supplemental irrigation is not provided. In the CTV treatment, GYs were slightly lower than GYs in the WW plot, but a significant difference was not seen. This indicates that plants in the CTV plot obtained sufficient water. Grain yields in the other temperature-scheduled plots were significantly reduced below those of the WW plot. Though, only about 50–60 mm of irrigation water was applied to these plots.

Global land surface warming is expected to cause extreme temperature events more frequently and more strongly, influencing the growth and development of the major cereal crops in a number of ways, thus affecting the production component of food security. Here, rice and maize crop responses to temperature in diverse, but constant, phenological phases and development stages were identified. A review of literature and data set of about 140 scientific articles have ascertained the crucial temperature thresholds and responses of rice and maize to extreme temperature, complementing a former study on wheat. Lethal temperatures and cardinal temperatures, along with error estimations, have been recognized for phenological phases and development stages. Temperature thresholds of the three crops for the important physiological processes such as initiation of the leaf, shoot growth and root growth, and for the most susceptible phenological phases, for example, sowing to emergence, anthesis, and grain filling were statistically analyzed by following the methodology of earlier work. The study summarized that the cardinal temperatures were conservative between researches and were apparently well described in all three crops. In all the three crops, anthesis and ripening were found to be the most temperature-sensitive stages (Sánchez et al., 2014).

7.2.3   ENZYMATIC ACTIVITIES UNDER HEAT STRESS

The average temperature in the U.S. Corn Belt during the grain-filling period of maize (*Zea mays* L.) is more than optimum for maximum GY. Wilhelm et al. (1999) determined the effects of a prolonged period of high temperature during grain filling on kernel growth (Fig. 7.3), composition, and starch metabolism of seven maize inbreds which were exposed to heat stress (33.5/25°C) or control (25/20°C) day/night temperature treatments in a greenhouse from 15 d after pollination (DAP) till maturity. They did not observe any considerable interaction that took place between genotype and temperature treatments for nine grain traits. There was a lengthening in the duration of grain filling by heat stress based on heat units. On a heat unit basis, they found that there was a decrease in kernel growth rate per heat unit. Further, there was also a dry weight loss of 7% in the mature kernel. Proportionally alike decreases were observed for starch, protein, and oil contents of the kernel. There was also a reduction in the kernel density.

FIGURE 7.3   Screening for heat stress tolerance at vegetative stage temperature effect on dent toughness.

Enzymes extracted from developing endosperm namely ADP glucose pyro-phosphorylase, glucokinase, sucrose synthase, and soluble starch synthase appeared to be highly sensitive to heat stress. However, when enzyme activities were adjusted with measured temperature coefficients (i.e., Q_{10}), only ADP glucose pyrophosphorylase showed reduced activity. These findings indicated that chronic heat stress during grain filling moderately inhibits seed storage processes and chooses starch metabolism enzymes to comparable degrees through multiple maize inbreds (Wilhelm et al., 1999).

Heat stress can reduce GYs and increase susceptibility to the aflatoxin-producing fungus, *Aspergillus flavus* in maize. Duke and Doehlert (1996) postulated that heat stress may affect gene expression in developing kernels, thus influencing enzyme activities and storage product accumulation. To verify this assumption, maize (*Zea mays* L. hybrid B73xMo17) kernels were cultured *in vitro* and grown at 25 (control), 30, and 35°C. At 5-day intervals, developing kernels were examined for levels of gene expression and enzyme activity. Message levels of all genes evaluated viz. *Shrunken-1, Shrunken-2, Brittle-2, Aldolase, Waxy, α-Zein, β-Zein,* and *Opaque-2* were decreased 65–80% after 5 days at 35°C. On the other hand, of the 17 enzymes assayed, only the activities of adenosine diphosphate glucose (ADP-Glc) pyrophosphorylase, aldolase, aspartate aminotransferase, acid invertase, and acid phosphatase were reduced by heat stress. Though the reduction in ADP-Glc pyrophosphorylase activity matched well with its decreased mRNA, aldolase activity roughly corresponded to its mRNA levels, and sucrose synthase activity was not affected by growth temperature, even though its transcript was decreased by 80%. They suggested that the reductions in enzyme activities in the heat-stressed kernels were due to declines in transcript. The different responses of enzyme activities to the heat stress may reflect variances in protein turnover. In the following experiment, they verified the assumption that heat stress may interfere with the ability of a maize kernel to produce pathogenesis response proteins. In cultured kernels grown at 25°C by treating with salicylic acid two-fold increment in the activity of the pathogenesis-related protein, chitinase, was observed. But, the salicylate-induced increment in chitinase activity was inhibited in kernels grown at 35°C. These findings suggested that heat stress prevents the induction of pathogenesis-related proteins in the maize kernel, and may, thus, interfere with the capability of the kernel to defeat pathogen invasion.

Lobell et al. (2013) reported that statistical studies of rainfed maize yields in the United States and other regions have showed two clear features:

a strong negative yield response to an increase of temperatures above 30 °C (or extreme degree days (EDD)), and a comparatively weak response to seasonal rainfall. They showed that the process-based Agricultural Production Systems Simulator (APSIM) is capable to reproduce both of these associations in the Midwestern United States. It provided insights into basic mechanisms. The principal effects of EDD in APSIM were associated with increased vapor pressure deficit, which causes water stress in two ways: by raising demand for soil water to maintain a given rate of carbon assimilation, and by decreasing future supply of soil water by increasing transpiration rates. Daily water stress as the ratio of water supply to demand was computed by APSIM and this ratio was found to be three times more responsive to 2°C warming than to a 20% reduction in precipitation during the critical month of July. The findings suggested a comparatively negligible role for direct heat stress on reproductive organs at present temperatures in this region. Effects of raised CO_2 on transpiration efficiency should lessen yield sensitivity to EDD in the next decades, but at most by 25%.

7.2.4   HEAT STRESS DURING REPRODUCTIVE GROWTH

Many researchers have indicated that in the future year's climate change is likely to occur, with frequent and severe occurrences of heat waves during most part of the year. Though several long-term studies on temperature stress were performed in maize, Siebers et al. (2017) analyzed the effects of immediate long-term effects of short duration extreme high temperature occurring during crucial developmental periods on physiological and yield parameters by application of infrared heating technology generated heat waves to field grown maize in east central Illinois. The heat waves generated from this infrared technology resulted in warming the canopy about 6°C above ambient canopy temperatures. These temperatures were maintained for three consecutive days during vegetative development (Wv1) and during an early reproductive stage (silking; Wv2) (Fig. 7.4). High temperatures given at both stages had no effect on the above ground vegetative biomass, but the heat stress at the silking stage had an effect in declining the reproductive biomass and drying of spikelets of tassel (Fig. 7.4). The high temperature treatment given at the early reproductive stage was found to bring about a reduction by 16% in the total reproductive biomass. This reduction in the reproductive biomass was due to a decline in the cob length as well as cob and husk mass. Though there was a decrease in the seed yield as well as KN, these reductions were not statistically significant. Heat stress resulted in transient reductions in leaf water potentials and midday photosynthesis, though there was no such impact on the soil water status. There was a complete recovery of leaf water potential and midday photosynthesis after the heat stress treatment; the results indicated that reproductive structures are more sensitive to the direct effects of high temperature stress rather than the vegetative structures.

FIGURE 7.4 Screening of maize for heat stress tolerance at reproductive stage-dried spikelets of tassel.

7.2.5 CHANGE IN GRAIN QUALITY BY HEAT STRESS

Heat stress occurring during the grain filling stage in maize has an effect on the physicochemical characteristics of the grain. The quality of the grain is affected. In four waxy maize varieties, the effect of heat stress (35.2/16.1°C with mean day and night temperatures) was imposed during the grain filling stage, that is, 1–40 days after pollination, was evaluated for its effect on the structure and thermal properties of starch accumulated in the grain by Lu et al. (2014). It was found that there was an increase in the granule size of the starch grains, with a proportionate increase in the amylopectin chains than the amylose chains. In addition, heat stress also resulted in the development of pitted starch granules of uneven surfaces. These starch granules had a declined swelling power. Though there was an increase in the gelatinization temperature and retrogradation percentage, there was no effect on the gelatinization enthalpy by heat stress imposed during the grain filling stage. The results thus indicated that heat stress prevailing during the grain filling stage of maize has an effect on the structural characteristics of waxy starch maize.

7.2.6   EXPRESSION OF HEAT SHOCK PROTEINS

Baszczynski et al. (1982) mentioned that exposure of 5-day-old plumules of corn (*Zea mays* L.) to raised temperatures for brief periods of time shifts the protein synthesis pattern from the production of a broad spectrum of proteins to the novel and (or) increased synthesis of a small number of heat shock polypeptides (HSPs). Most prominent is the low synthesis of a major polypeptide (relative mass (Mr) = 93,000 and isoelectric point = 8.0) normally made at 27°C and the increased and (or) new synthesis of polypeptides with Mrs of 108,000; 89,000; 84,000; 76,000; 73,000; and 18,000, following 1 h of heat shock. These six HSPs resolve into 18 spots by two-dimensional fluorographic analysis. The induction of the HSPs needs temperatures at or above 35°C for noticeable synthesis. Some of the HSPs are synthesized following only 15 min at 41°C and synthesis of all HSPs is observed within 120 min following heat shock. Recovery from heat shock is fast; after 6 to 8 h at 27°C following heat shock, the polypeptide pattern is identical with the control. Polypeptide synthetic patterns produced from the extracts of individual heat shocked shoots were indistinguishable to those from extracts of 20 shoots, irrespective of whether single shoots were intact or excised during labeling. Polypeptide synthetic patterns and response of single 5-day-old primary roots to heat shock were identical to shoots. This was the first demonstration of the heat shock polypeptides induction in a whole, intact higher plant.

Dupuis and Dumas (1990) investigated the response of maize (*Zea mays*) male and female mature reproductive tissues to temperature stress. The fertilization rate was more declined when pollinated spikelets were subjected to temperatures above 36°C. When pollen and spikelets were subjected to temperature stress separately, the female tissues seem resistant to 4 h of cold stress (4°C) or heat stress (40°C). Under heat shock conditions, the production of a typical set of HSPs is induced in the female tissues. On the other hand, the mature pollen was sensitive to heat stress and was accountable for the failure of fertilization at high temperatures. Heat shock response was found in the mature pollen at the molecular level.

7.2.6.1 COLD STRESS

Massonneau et al. (2005) carried out inclusive searches of maize EST data which made possible the identification of 8 new *Corn Cystatin* (*CC*) genes besides the earlier known genes *CCI* and *CCII*. They found that in all 10 genes the assumed amino acid sequences had typical cystatin family signature. Furthermore, they show a prolonged general similarity with cystatins from other species belonging to a number of diverse phytocystatin subfamilies. To get an extra understanding of their respective roles in the maize plant, gene-specific expression profiles were created by semi-quantitative RT-PCR. Though 7 *CC* genes were expressed in two or more tissues differing from gene to gene, *CCI* was preferentially expressed in immature tassels and *CC8* and *CC10* in developing kernels. The *in situ* hybridization of maize kernels revealed that *CC8* was specifically expressed in the basal region of the endosperm and *CC10* both in the starchy endosperm and the scutellum of the embryo. The other non-kernel-specific genes had different expression kinetics during kernel development, normally with peaks during the early stages. Along with developmental regulation, the effect of cold stress and water starvation was verified on cystatin expression. The cold stress induced two genes (*CC8* and *CC9*) and 5 genes (*CCII, CC3, CC4, CC5,* and *CC9*) were down-regulated in response to water starvation. Altogether this data suggested diverse functions for *CC* genes in the maize plant.

7.2.6.2 CHILLING STRESS

An investigation of Louarn et al. (2010) on the effect of transient chilling in two genotypes of maize has shown that chilling temperatures did have an impact on the kinetics of all emerging leaves. However, these chilling temperatures did not have an effect on the final leaf length of these genotypes. They did not observe any size mediated propagation of reduction of an initial growth; however, it was found that a longer duration of leaf elongation was responsible for compensating the declined rates of leaf elongation during their initial or early stages of growth. Cold-induced response at leaf level in maize is contributed by the cell division and cell expansion rates. Thus, their findings have demonstrated that in maize the kinetics of leaf elongation and the attainment of final leaf length are under the control of processes at the n − 1 (cell proliferation and expansion) and n + 1 (whorl size signal) scales. Both levels may respond to chilling stress with different time lags, making it possible to buffer short-term responses.

7.2.7   QUANTITATIVE TRAIT LOCI CONTROLLING COLD-RELATED TRAITS

The enhancement of early vigour is essential for the adaptation of maize (*Zea mays* L.) to the climatic conditions of central Europe and the northern Mediterranean. There early sowing is a key strategy for evading the effect of summer drought. Hund et al. (2004) found the QTL regulating cold-related characters. A set of 168 $F_{2:4}$ families obtained from the cross Lo964 × Lo1016 was raised in a sand–vermiculite substrate at 15/13°C (day/night) till the one-leaf stage. Twenty QTL were identified for the four shoot and two seed traits studied.

The study of root weight and digital measurements of the length and diameter of primary and seminal roots helped in the detection of 40 QTL. The operating efficiency of photosystem II (Φ_{PSII}) was associated with seedling dry weight at both the phenotypic and genetic level ($r = 0.46$, two matching loci, respectively) but it did not show any relation with root characteristics. Cluster analysis and QTL association unveiled that the various root characteristics were largely independently inherited and that root lengths and diameters were mainly negatively correlated. In this study, the major QTL for root traits identified in a previous study in hydroponics were established. The length of the primary lateral roots showed a negative association with the germination index ($r = -0.38$). A previous study in hydroponics established a large number of independently inherited loci appropriate for the enhancement of early seedling growth through better seed vigor and/or a higher rate of photosynthesis.

Variations in anthocyanin content and transcript abundance for genes whose products function in general phenylpropanoid metabolism and the anthocyanin pathway were examined in maize (*Zea mays* L.) seedlings during the short-term, low-temperature treatment by Christie et al. (1994). With the increase in severity and duration of cold stress, an increment in anthocyanin and mRNA abundance was observed in sheaths of maize seedlings. In tested lines capable of producing anthocyanin geno typically, anthocyanin accumulation was noticed. Within 24 h after transfer of 7-d maize (B37N) seedlings to 10° C, phenylalanine ammonia-lyase (*Pal*) (EC 4.3.1.5)-homologous and chalcone synthase (*C2*) (EC 2.3.1.74) transcript levels increased by 8- and 50-fold, respectively, and 4-coumarate: CoA ligase (*4Cl*) (EC 6.2.1.12)-homologous and chalcone isomerase (*Chi*) (EC 5.5.1.6)-homologous transcripts increased by 3-fold over levels in unstressed plants. Time-course studies revealed that *Pal* (EC 4.3.1.5) and *C2*-transcript levels remained constant for the first 12 h of cold stress, then increased markedly over the next 12 h, and dropped to pretreatment levels within 2 d of returning cold stressed seedlings to ambient (25° C) tempera-ture. Within 6 h of cold stress transcripts *4Cl* (EC 6.2.1.12) and *Chi* (EC 5.5.1.6) increased considerably, then further increase was not observed over the next 36 h, and dropped to pretreatment levels upon returning seedlings to 25° C. During cold treatment, transcripts homologous to two regulatory (*R, C1*) and three structural (*A1, A2,* and *Bz2*) anthocyanin genes increased by 7- to 10-fold, showing similar kinetics of accumulation as for *Pal* (EC 4.3.1.5) and *C2* transcripts. Transcripts encoded by *Bz1*, the anthocyanin structural gene for UDP: glucose-flavonol glucosyltransferase (EC 2.4.1.91), were comparatively abundant in control tissues and showed only a transient increase during the cold period. This study suggested that the genes of the anthocyanin biosynthetic pathway can be deliberated *c or* (Cold-Regulation) genes, and as this pathway is well described, it is an outstanding subject for characterizing plant molecular responses to low temperatures.

Most plant oxylipins are made through the lipoxygenase (LOX) pathway. Current advancement in dicots has brought to light the biological roles of oxylipins in plant defense responses to pathogens and pests. On the other hand, the physiological function of LOXs and their metabolites in monocots is not well understood. Two maize LOXs pathways *ZmLOX10* and *ZmLOX11* that have >90% amino acid sequences in identical but are located on different chromosomes were cloned and characterized by

Nemchenko et al. (2006). Phylogenetic analysis unveiled that *ZmLOX10* and *ZmLOX11* cluster together with well-characterized plastid ic *type 2* linoleate 13-LOXs from different plant species. Regio-specificity analysis of recombinant ZmLOX10 protein over expressed in *Escherichia coli* showed it to be a linoleate 13-LOX with a pH optimum at ~pH 8.0. Both predicted proteins have putative transit peptides for chloroplast import. *ZmLOX10* was especially expressed in leaves and was induced in response to wounding, cold stress, defense-related hormones jasmonic acid (JA), salicylic acid (SA), and abscisic acid (ABA), and inoculation with an avirulent strain of *Cochliobolus carbonum*. These findings indicated a role of this gene in maize adaptation to abiotic stresses and defense responses against pathogens and pests. *ZmLOX11* was especially expressed in silks and was induced in leaves only by ABA, signifying its possible contribution in responses to osmotic stress. In leaves, mRNA accumulation of *ZmLOX10* is strictly regulated by a circadian rhythm with maximal expression coinciding temporally with the highest photosynthetic activity. Their research unveiled the evolutionary divergence of physiological roles for comparatively newly duplicated genes and the physiological functions of these 13-LOXs.

Four species of protein kinase were found in senescent maize leaves by means of a gel assay for kinase activity with myelin basic protein (MBP) as the substrate by Berberich et al. (1999). In healthy green leaves that had been subjected to low-temperature stress (5°C) and then returned to 25°C most of these kinases were identified. A 39-kDa protein was activated by cold stress. The other two proteins, of 35 kDa and 52 kDa, constitutively phosphorylated MBP during senescence and temperature up-shift. Judging from their molecular masses, cation requirements, and substrate specificities, it appeared likely that the 39-kDa, 41-kDa, and 45-kDa proteins signified mitogen-activated protein kinases (MAPKs). Consequently, two MAPK cDNAs were isolated from a cDNA library constructed using mRNAs from senescent leaves. Northern analysis exhibited that the transcript corresponding to one of the cDNAs, designated *ZmMPK5*, collected in healthy leaves 3 h after the up-shift to 25°C as well as in senescent leaves, suggesting that the *ZmMPK5* encodes 45-kDa protein kinase. Western analysis using an antiserum against the C-terminal region of ZmMPK5 showed that the level of the *ZmMPK5* protein increased in senescent leaves. These findings suggested that a 45-kDa MAPK is involved in the senescence process and recovery from low-temperature stress in maize.

7.3   FLOODING/HIGH MOISTURE STRESS

Zaidi et al. (2003) conducted experiments in six genotypes for their responses to waterlogging stress. They raised the seedlings in disposable plastic cups (250 cm^3) under complete saturated soil conditions and transplanted them in the field 20 days after planting and again subjected to waterlogging stress at knee-high stage for 7 days. The growth, biochemical composition, and metabolic activities were severely affected with ESM stress both at the early stage and knee-high stage. The genotypes exhibited a similar response to ESM stress at two growth stages. Genotypes with early adventitious rooting, partial stomatal closure, <5.0 days anthesis-silking interval, increased root NAD$^+$-alcohol dehydrogenase activity and high starch accumulation in stem tissues exhibited good tolerance to ESM stress. Both preexisting and induced (pre-hypoxia/anoxia) tolerance had most of these morphophysiological traits related with ESM tolerance in common. The study results revealed that hypoxia/anoxia pretreatment improves tolerance to waterlogging conditions in maize.

Excess moisture (waterlogging) during the summer–rainy season is a major limitation in maize production (*Zea mays* L.) in many regions of Southeast Asia. Selection and development of genotypes tolerant to stress conditions could be an affordable method appropriate for resource poor maize-growing farmers of such areas. Zaidi et al. (2004) make an effort to find the most susceptible/critical crop stage(s) of maize for excess moisture stress. Among the four crop stages, viz. early seedling (V2), knee-high (V7), tasseling (VT) and milk stage (R1), early seedling was found to be highly susceptible, followed by the tasseling stage. A screening technique (cup method) was developed/standardized and was found to be an effectual technique for large-scale screening of maize genotypes against excess soil moisture stress (Fig. 7.5). Germplasm was screened using this technique and then field evaluation was performed at the V7 growth stage (seventh leaf visible). Various growth and biochemical parameters were affected severely with excess soil moisture stress; it impaired anthesis and silking, and finally resulted in poor development of kernel and yield. However, significant variability was observed among the genotypes screened. Genotypes with good accumulation of carbohydrate in stem tissues, moderate stomatal conductance, <5 days ASI, high root porosity, and early brace root development ability exhibited high tolerance against the hypoxia/anoxia caused by excess soil moisture conditions.

FIGURE 7.5   Screening of maize germplasm for flooding tolerance.

Saab and Sachs (1996) isolated a flooding-induced maize (*Zea mays* L.) gene (wusl1005 [gfu]; abbreviated as 1005) encoding a homolog of xyloglucan endo-transglycosylase (XET), a putative cell-wall loosening enzyme active during germination, expansion, and fruit softening. XET and its related enzymes may also be involved in cell-wall metabolism during flooding-induced aerenchyma development. Under flooded conditions, 1005 mRNA accumulates in root and mesocotyl locations and develops aerenchyma, which reaches maximum levels within 12 h of treatment. Development of aerenchyma was noticed in the same locations by 48 h of treatment. In both root and mesocotyl, the treatment with the ethylene synthesis inhibitor (aminooxy) acetic acid (AOA), prevented the formation of cortical air space and completely inhibits 1005 mRNA accumulation under flooding. AOA treatment exhibited little effect on the mRNA accumulation encoded by adh1 showing that it did not cause overall suppression of flooding-responsive genes. Moreover, ethylene treatment under aerobic conditions brings about aerenchyma development along with 1005 induction in both organs. These findings specify that 1005 is responsive to ethylene. Treatment with anoxia, which overwhelms ethylene accumulation and aerenchyma development, also caused 1005 induction. But, in contrast to flooding, AOA treatment under anoxia did not affect 1005 mRNA accumulation, showing that 1005 is induced through various mechanisms under flooding (hypoxia) and anoxia (Saab and Sachs, 1996).

High water table conditions decrease crop yields. Evans et al. (1991) developed corn and soybean relative yield models for high water table conditions. The relative yield models were based on stress-day-index (SDI) associations using SEW30 (0.3 m water table depth) to explain the high water table stress criteria and normalized SDI crop susceptibility (CS) factors. The normalized crop susceptibility (NCS) factors were ascertained from earlier studies carried out in North Carolina. The models were developed with obtainable field data for SDI and crop yield from Ohio. The resultant com model was tested against data from India and North Carolina and elucidated 69% of the comparative yield variance for the pooled data. The soybean model elucidated 66% of the variance in relative yield for 6 years of soybean data from Ohio. The models developed should improve relative yield estimations using DRAINMOD, a water table management simulation model.

Flooding affects plant growth. Hwang and VanToai (1991) have revealed that treatment of corn (*Zea mays* L.) seedlings with ABA improves their tolerance to anoxia 10-fold. Corn seedlings planted in vermiculite and exposed to stress anoxically for 1 day had only 8% survival. Root tips pretreated with 100 micromolar ABA or water for 24 h before the 1 day anoxic stress exhibits increment in anoxic survivability of seedlings to 87 and 47%, respectively. The addition of cycloheximide (5 mg per liter) along with ABA decreased the seedling survival rate, signifying that the induction of anoxic tolerance in corn by ABA was partially a result of the new protein synthesis. ABA treatment brings about a threefold increase in alcohol dehydrogenase enzyme activity in corn roots. But, after 24 h of anoxia, a significant difference was not noticed in alcohol dehydrogenase enzyme activity of the ABA-pretreated and non-pretreated corn roots. The results showed that ABA plays a significant role in inducing anoxic tolerance (Fig. 7.6) in corn and that the induced tolerance was most likely mediated by an increase in alcohol dehydrogenase enzyme activity before the anoxic stress.

Excessive moisture (EM) stress during the summer–rainy season is a major production constraint for maize (*Zea mays* L.) in many regions of South and South-East Asia. Zaidi et al. (2007) attempted to find the association between morphophysiological characters of flooding tolerant line and susceptible line (Fig. 7.7) and GY measured on inbred parents and their single cross progenies under EM stress. Responses of different morphophysiological traits, except days to 50% anthesis, differed considerably. The superiority of hybrid progenies over parental inbred lines was elevated under EM stress. This suggested that these hybrids were more tolerant to EM stress than inbred progenies. Across moisture regimes, all morphophysiological traits of hybrids, except lodging and root porosity under normal moisture, showed a positive and significant correlation with midparent traits. Therefore, their study data has proposed that *per se* performance of lines was comparatively an additional key factor in defining hybrid performance under EM stress, while under optimum soil moisture conditions midparent heterosis was comparatively more important than *per se* performance of midparent. Phenotypic correlation between hybrid and midparent yields exhibited a strong association under EM stress ($r = 0.66^{**}$). The association was statistically significant under normal moisture as well, though it was relatively weak ($r = 0.41^{*}$). Their results suggested that under EM, the performance of hybrid progenies can be predicted and enhanced to some extent based on their inbred parents that have been scientifically selected and improved for EM stress.

FIGURE 7.6   Flooding tolerance: tolerant lines not affected and susceptible lines (rotting of culm and roots in susceptible).

The cell wall hydrolysis and subsequent cell lysis develop aerenchyma (soft cortical tissue with large intercellular air spaces) in flooded plants, which is promoted by endogenous ethylene (Fig. 7.8). In spite of its adaptive importance, the molecular mechanisms behind the development of aerenchyma remain unknown.

FIGURE 7.7 Under flooded conditions: variation in growth between tolerant line (left side) and susceptible line (right side).

Maize flooding Control line T.S. of Root (10x, 40x)

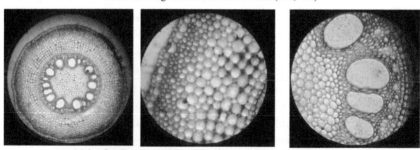

Maize flooding Treated line T.S. of Root (10x, 40x)

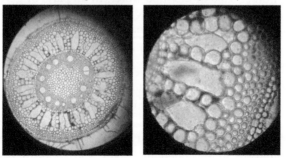

FIGURE 7.8 Formation of aerenchyma in root transverse section in tolerant lines.

7.4 SALT STRESS

The effect of NaCl and Na_2SO_4 on the growth of two maize cultivars (*Zea mayscv.* Pioneer 3906 and cv. Across 8023) varying in Na^+ uptake was investigated in two greenhouse experiments by Fortmeier and Schubert (1995). Na^+ treatment with different accompanying anions (Cl^-/SO_4^{2-}) revealed that Na^+ causes ion toxicity. During the first 2–3 weeks, the shoot growth of the two cultivars was more affected. There was also a reduction in the shoot Ca^{2+} concentration. The youngest leaves had much lower concentration. During this first phase, higher concentrations of Ca^{2+} were noted in Across 8023 than Pioneer 3906. The Na^+ excluding cultivar Pioneer 3906 exhibited continuous, though reduced, growth compared with the control. There was a decline in the Na^+ concentration in the shoot up to flowering. Cultivar Across 8023 accumulated Na^+ till flowering. The growth of stressed plants was much reduced than that for Pioneer 3906. Clear toxic symptoms were observed on leaves of cultivar Across 8023, while those of the more salt tolerant cultivar Pioneer 3906 did not. It was resolved that Na^+ exclusion contributes to the salt tolerance of maize.

Ashraf and Rauf (2001) conducted a study to find out whether salt tolerance could be induced in maize at the germination stage by soaking of seeds for 8 h in distilled water or 200 meq·L^{-1} of NaCl, KCl, $CaCl_2$•$2H_2O$. Both primed and unprimed seeds were exposed to 0, 100, or 200 mol·m^{-3} NaCl for 14 days under controlled conditions. Though all priming agents effectively alleviated harmful effects of salt stress on maize at the germination stage, $CaCl_2$•$2H_2O$ was found to be more effective as significantly higher final germination rate, fresh and dry weights of plumules and radicals were found in the seeds primed with this salt than those treated with other salts or distilled water. In all parts of germinating seeds of maize seeds primed with NaCl, KCl, or $CaCl_2$•$2H_2O$ significantly higher concentration of Na^+, K^+, and Ca^{2+} was recorded, respectively. Additionally, seeds primed with $CaCl_2$•$2H_2O$ showed the maximum accumulation of Cl^- in all parts of the germinating seeds, followed by seeds treated with NaCl and KCl. Maximum Ca^{2+} was retained in seeds and mesocotyl, due to which, transport of this ion to plumules and radicals was low.

The effect of salt stress on the activity of antioxidative enzymes and lipid peroxidation were investigated in leaves and roots of two maize genotypes, BR5033 (salt-tolerant) and BR5011 (salt-sensitive), grown under control (nutrient solution) or salt stress (nutrient solution containing 100 mM NaCl) conditions by De Azevedo Neto et al. (2006). In leaves of salt-stressed plants, superoxide dismutase (SOD), ascorbate peroxidase (APX), guaiacol peroxidase (GPX), and glutathione reductase (GR) activities increased with time when compared to the controls. This increase in enzyme activities was more marked in the salt-tolerant genotypes. Salt stress did not have much significant effect on catalase (CAT) activity in the salt-tolerant genotype. In salt-stressed roots of the salt-tolerant genotype, SOD and CAT activities were reduced and APX, GPX, and GR activities remained unchanged. In the roots of the salt-sensitive genotype, salinity led to a reduction in the activity of the above enzymes. In both leaves and roots, CAT and GPX enzymes showed the maximum H_2O_2 scavenger activity. Furthermore, CAT, APX, and GPX activities in conjunction with SOD seemed to play a significant protective role in the scavenging processes. Lipid peroxidation was improved only in salt-stressed leaves of the salt-sensitive genotype. Their findings indicated that oxidative stress may perform a significant role in salt-stressed maize plants and that the greater protection of BR5033 leaves and roots from salt-induced oxidative damage results, at least in part, because of the maintenance and/or increase of the antioxidant enzymes activity.

An endogenous signaling molecule, nitric oxide (NO) facilitates responses to abiotic and biotic stresses in animals and plants. Zhang et al. (2006) investigated how NO regulates Na^+, K^+ ion homeostasis in maize. Pretreatment with 100 µm SNP for 2 days enhanced later growth of maize plants under 100 mMNaCl stress, which was showed by improved dry matter accumulation, increased chlorophyll content, and reduced membrane leakage from leaf cells. An NO scavenger, methylene blue (MB-1), blocked the effect of SNP. The study findings showed that SNP-derived NO improved maize tolerance to salt stress. Additional analysis revealed that NaCl induced a transient increase in the NO level in maize leaves. Both NO and NaCl treatment stimulated vacuolar H^+-ATPase and H^+-PPase activities, leading to increased H^+-translocation and Na^+/H^+ exchange. NaCl-induced H^+-ATPase and H^+-PPase activities were reduced by MB-1. 1-Butanol, an inhibitor of phosphatidic acid (PA) production by phospholipase D (PLD), reduced NaCl- and NO-induced H^+-ATPase activation. On the other hand, applied PA stimulated H^+-ATPase activity. Therefore, the study suggested that NO acts as a signal molecule in the NaCl response by rising the activities of vacuolar H^+-ATPase and

H$^+$-PPase, which offers the driving force for Na$^+$/H$^+$ exchange. PLD and PA play a significant role in this process.

Maize (*Zea mays L.*) plants in the initial stage of development were treated with 80 mM sodium chloride (NaCl) with or without supplemental calcium (Ca^{2+}) (8.75 mM) for 7 days by Alberico and Cramer (1993).

The influence of salinity on dry matter production and shoot and root concentrations of sodium (Na$^+$), Ca^{2+}, and potassium (K$^+$) were evaluated in seven Pioneer maize cultivars. Salinity decreased total dry weight, leaf area and shoot and root dry weight. In all seven cultivars, reduced Na$^+$ concentrations and a significant increase in leaf area were observed with the supplement of salinized nutrient solutions with 8.75 mM calcium chloride (CaCl$_2$). Under NaCl salinity, the two cultivars with the lowest shoot and root Na$^+$ concentrations exhibited the maximum increment in total, shoot and root dry weights with the addition of supplemental Ca. Significant reduction in shoot fresh weight/dry weight ratios was observed in all cultivars by both salinity treatments, but supplemental Ca^{2+} increased the ratio relative to salinity treatments without supplemental Ca. Root fresh weight/dry weight ratios were reduced only by salinity treatments with supplemental Ca. With NaCl salinity, cultivars with lower shoot and root Na$^+$ concentrations were found to be more salt sensitive and had significantly lower amounts of dry matter production when compared with cultivars having higher shoot and root Na$^+$ concentrations. It was established that Na$^+$ exclusion from the shoot was not linked with and was an inaccurate indicator of salt tolerance for maize.

Salicylic acid effects on some physiological and biochemical characteristics of maize (*Zea mays* L.) seedlings under NaCl stress were investigated by Gautman and Singh (2009). Maize seeds were given presoaking treatments of NaCl (0, 50, 100, and 200 mM) in both the presence and absence of 0.5 mM salicylic acid. Two-week-old maize seedlings exhibited significant decreases in dry weight, root length, shoot length, and leaf area on 6 h exposure of 100 and 200 mMNaCl stress. Photosynthetic pigments and NR activity in leaves declined sharply with increasing stress levels. Under saline conditions, an increment was observed in both proline content and lipid peroxidation (measured in terms of MDA) levels. But, seedlings pretreated with 0.5 mM salicylic acid together with the salinity levels have shown improvement in growth parameters, photosynthetic pigments, NR activity whereas, free proline and MDA levels got reduced. The results showed that salt-induced deleterious effect in maize seedlings was significantly encountered by the pretreatment of salicylic acid. It is resolved that 0.5 mM salicylic acid increases the adaptability of maize plants to NaCl stress.

Cramer et al. (1994) tested the proposition that accumulation of Na^+ in the shoot in maize has a negative association with salt tolerance. Salt tolerance is described as a percentage of the control on a dry weight basis. Two hybrids (Pioneer hybrid 3578 and Pioneer hybrid 3772) varying extensively in Na^+ accumulation were compared. Plants were treated with two types of salinity for 15 days (80 mol m^{-3} NaCl or 80 mol m^{-3} NaCl plus 8.75 mol m^{-3} $CaCl_2$). Ion concentrations (Na^+, K^+, Ca^{2+}, and Cl^-) were measured in the roots, stalks, sheaths, and leaves of plants harvested every third day. Ion concentrations were significantly influenced by the treatments. Salinity treatments increased the Na^+ and Cl^- concentrations and K^+ and Ca^{2+} concentrations. Further increase of Ca^{2+} reduced Na^+ concentrations. Hybrid 3772 retained very low Na^+ concentrations in the shoots, whereas 3578 did not. The ability to transport Na^+ to the shoot was the major difference between the hybrids hybrid; 3578 transported Na^+ at twice the rate of hybrid 3772. Generally, ion transport to the shoot seemed to be a function of root ion concentration. This model could account for the effects of NaCl salinity and supplemental Ca^{2+} onion transport, though Na^+ transport was made complex by an apparent reabsorption mechanism in the root and mesocotyl. The absence of association of Na^+ accumulation in the shoot and other ion parameters with growth showed that the mineral nutrition of the plants was not associated with salt tolerance. It was determined that the growth response of maize to salinity was mainly affected by osmotic factors.

7.4.1 REDUCTION IN KERNEL SETTING UNDER SALT STRESS

Salt stress has its influence on the grain yields (GYs). The yields of maize under salt stress may exhibit a decline due to a reduction in the kernel setting rather than declined grain filling. Hütsch et al. (2014) tested whether there was a decrease in the acid invertase activity in developing kernels of maize, if there was a limitation in the supply of assimilates under salt stress, in a salt-sensitive maize hybrid, Pioneer 3906 (relatively salt-sensitive 0 and a hybrid SR 12 (moderately salt tolerant). The findings showed that there was a significant decline in the GYs, which were the resultant of a

50% decline in the kernel number. However, it was found that in both these hybrids there was hardly any limitation in source, where the concentrations of sucrose in the kernels of both these hybrids were found at increased levels even under salt stress. In contrast to the increased concentrations of sucrose in the kernels, the concentrations of glucose and fructose were lower. A considerable decline (19%) in the soluble acid invertase activity was seen in kernels 5 days after pollination under hydroponically grown plants, while there was a reduction of 50% in the activity in kernels of those grown under soil culture experiment at 2 days after pollination. Though the decline in the enzyme activity was identical in both the sensitive as well as moderately tolerant hybrids, the highest soluble acid invertase activity that resulted at 2 days after pollination exhibited a sharp decline until 8 days after pollination in soil grown plants in Pioneer 3906. Thus, the results determined that a reduction in the acid invertase activity is the main factor related to the limitation of kernel setting apart from other genotypic variabilities.

7.5   HEAVY METAL STRESS

Heavy metals like cadmium (Cd) enter the ecosystem largely due to human activities. Currently, the heavy metal pollution of the soil is prod ucing ever greater problems, intensified by the fact that the heavy metals accumulated in plants may, either directly or indirectly, enter into animals and human beings. Maize is one among the global important crops, standing third after wheat and rice, so the variations made by one of the most toxic heavy metals, Cd, in maize plants is a matter of some concern. Pál et al. (2006) reviewed the toxic symptoms caused by Cd stress, and the tolerance mechanisms activated in the plants. Cadmium induces some physiological variations, such as growth inhibition, variations in the water and ion metabolism, the inhibition of photosynthesis, variations in enzyme activities, and the formation of free radicals. The initiation of Cd stress induced the synthesis and compartmentalization of phytochelatins (Pál et al., 2006).

7.6   CONCLUSIONS

Research studies undertaken to understand the effects of different abiotic stresses have been discussed. Abiotic stresses caused by temperature and water are being intensified as a result of climate change. To overwhelm this problem and meet the worldwide demand for food, tolerant crops must be developed using suitable technologies. Physiological characteristics, tolerance mechanisms, and management strategies for better crop production must be investigated in detail to efficiently apply such technologies.

KEYWORDS

- maize
- abiotic stress
- temperature
- water
- research advances

REFERENCES

Alberico, G. J.; Cramer, G. R. Is Salt Tolerance of Maize Related to Sodium Exclusion?. Preliminary Screening of Seven Cultivars. *J. Plant Nutr.* **1993**, *16*, 2289–2303.

Amin, M.; Ahmad, R.; Ali, A.; Hussain, I.; Mahmood, R.; Aslam, M.; Lee, D. J. Influence of Silicon Fertilization on Maize Performance Under Limited Water Supply. *Silicon* **2018**, *10*, 177–183.

Ashraf, M.; Rauf, H. Inducing Salt Tolerance in Maize (*Zea mays* L.) Through Seed Priming with Chloride Salts: Growth and Ion Transport at Early Growth Stages. *Acta Physiol. Plant.* **2001**, *23*, 407–414.

Bahadori, A.; Mobasser, H.; Reza, H. Influence of Water Stress and Plant Density on Some Characteristics in Corn. *Biol. Forum Int. J.* **2015**, 673–678.

Banziger, M.; Edmeades, G. O.; Lafitte, H. R. Physiological Mechanisms Contributing to the Increased N Stress Tolerance of Tropical Maize Selected for Drought Tolerance. *Field Crops Res.* **2002**, *75*, 223–233.

Berberich, T.; Sano, H.; Kusano, T. Involvement of a MAP Kinase, ZmMPK5, in Senescence and Recovery from Low-Temperature Stress in Maize. *Mol. Gen. Genet.* **1999**, *262*, 534–542.

Baszczynski, C. L.; Walden, D. B.; Atkinson, B. G. Regulation of Gene Expression in Corn (*Zea mays* L.) by Heat Shock. *Can. J. Biochem.* **1982**, *60*, 569–579.

Bruce, W. B.; Edmeades, G. O.; Barker, T. C. Molecular and Physiological Approaches to Maize Improvement for Drought Tolerance. *J. Exp. Bot.* **2002**, *53*, 13–25.

Clawson, K. L.; Blad, B. L. Infrared Thermometry for Scheduling Irrigation of Corn. *Agron. J.* **1982,** *74*, 311–316.

Chimungu, J. G.; Brown, K. M.; Lynch, J. P. Reduced Root Cortical Cell File Number Improves Drought Tolerance in Maize. *Plant Physiol.* **2014,** *166*, 1943–1955.

Christie, P. J.; Alfenito, M. R.; Walbot, V. Impact of Low-Temperature Stress on General Phenylpropanoid and Anthocyanin Pathways: Enhancement of Transcript Abundance and Anthocyanin Pigmentation in Maize Seedlings. *Planta* **1994,** *194*, 541–549.

Cramer, G. R.; Alberico, G. J.; Schmidt, C. Salt Tolerance is Not Associated with the Sodium Accumulation of Two Maize Hybrids. *Aust. J. Plant Physiol.* **1994,** *21*, 675–692.

De Azevedo Neto, A. D.; Prisco, J. T.; Enéas-Filho, J.; Braga de Abreu, C. E.; Gomes-Filho, E. Effect of Salt Stress on Antioxidative Enzymes and Lipid Peroxidation in Leaves and Roots of Salt-Tolerant and Salt-Sensitive Maize Genotypes. *Environ. Exp. Bot.* **2006,** *56*, 87–94.

Duke, E. R; Doehlert, D. C. Effects of Heat Stress on Enzyme Activities and Transcript Levels in Developing Maize Kernels Grown in Culture. *Environ. Exp. Bot.* **1996,** *36*, 199–208.

Dupuis, I.; Dumas, C. Influence of Temperature Stress on *in Vitro* Fertilization and Heat Shock Protein Synthesis in Maize (*Zea mays* L.) Reproductive Tissues. *Plant Physiol.* **1990,** *94*, 665–670.

Evans, R. O.; Skaggs, R. W.; Sneed, R. E. Stress Day Index Models to Predict Corn and Soybean Relative Yield Under High Water Table Conditions. *Trans. ASAE.* **1991,** *34*, 1997–2005.

Fischer, K. S.; Edmeades, G. O.; Johnson, E. C. Selection for the Improvement of Maize Yield Under Moisture-Deficits. *Filed Crops Res.* **1989,** *22*, 227–243.

Fortmeier, R.; Schubert, S. Salt Tolerance of Maize (*Zea mays* L.): The Role of Sodium Exclusion. *Plant Cell Environ.* **1995,** *18*, 1041–1047.

Gautam, S.; Singh, P. K. Salicylic Acid-Induced Salinity Tolerance in Corn Grown Under NaCl Stress. *Acta Physiol. Plant.* **2009,** *31*, 1185–1190.

Heckathorn, S. A.; Poeller, G. J.; Coleman, J. S.; Hallberg, R. L. Nitrogen Availability Alters Patterns of Accumulation of heat stress-induced Proteins in Plants. *Oecologia* **1996,** *105*, 413–418.

Hund, A.; Fracheboud, Y.; Soldati, A.; Frascaroli, E.; Salvi, S.; Stamp, P. QTL Controlling Root and Shoot Traits of Maize Seedlings Under Cold Stress. *Theor. Appl. Genet.* **2004,** *109*, 618–629.

Hütsch, B. W.; Saqib, M.; Osthushenrich, T.; Schubert, S. Invertase Activity Limits Grain Yield of Maize Under Salt Stress. *J. Plant Nutr. Soil Sci.* **2014,** *177*, 278–286.

Hwang, S. Y.; VanToai, T. T. Abscisic Acid Induces Anaerobiosis Tolerance in Corn. *Plant Physiol.* **1991,** *97*, 593–597. DOI: https://doi.org/10.1104/pp.97.2.593.

Li, R.; Zeng, Y.; Xu, J.; Wang, Q.; Wu, F.; Cao, M.; Lan, H.; Liu, Y.; Lu, Y. Genetic Variation for Maize Root Architecture in Response to Drought Stress at the Seedling Stage. *Breed. Sci.* **2015,** *65*, 298–307.

Lobell, D. B.; Hammer, G. L.; McLean, G.; Messina, C.; Roberts, M. J.; Schlenker, W. The Critical Role of Extreme Heat for Maize Production in the United States. *Nat. Clim. Change.* **2013,** *3*, 497–501.

Louarn, G.; Andrieu, B.; Giauffret, C. A Size-Mediated Effect can Compensate for Transient Chilling Stress Affecting Maize (*Zea mays*) Leaf Extension. *New Phytol.* **2010,** *187*, 106–118.

Lu, D.; Shen, X.; Cai, X.; Yan, F.; Lu, W.; Shi, Y. C. Effects of Heat Stress During Grain Filling on the Structure and Thermal Properties of Waxy Maize Starch. *Food Chem.* **2014,** *143,* 313–318.

Manavalan, L. P.; Musket, T.; Nguyen, H. T. Natural Genetic Variation for Root Traits Among Diversity Lines of Maize (*Zea mays* L.). *Maydica* **2011,** *56,* 1–10.

Massonneau, A.; Condamine, P.; Wisniewski, J. P.; Zivy, M.; Rogowsky, P. M. Maize Cystatins Respond to Developmental Cues, Cold Stress and Drought. *Biochim. Biophys. Acta* **2005,** *1729,* 186–199.

Mize, C. W.; Egehand, M. H.; Batchelor, W. D. Predicting Maize and Soybean Production in a Sheltered Field in the Cornbelt Region of North Central USA. *Agrofor. Syst.* **2005,** *64,* 107–116.

Muchow, R. C. Comparative Productivity of Maize, Sorghum and Pearl Millet in a Semi-Arid Tropical Environment II. Effect of Water Deficits. *Field Crops Res.* **1989,** *20,* 207–219.

Nelsaon, D. E.; Repetti, P. P.; Adams, T. R.; Creelman, R. A.; Wu, J.; Warner, D. C.; Anstrom, D. C.; Bensen, R. J.; Castiglioni, P. P.; Donnarummo, M. G.; Hinchey, B. S.; Kumimoto, R. W.; Mazle, D. R.; Canales, R. D.; Krolikowski, K. A.; Dotson, S. B.; Gutterson, N.; Ratcliffe, O. J.; Heard, J. E. Plant Nuclear Factor Y (NF-Y) B Subunits Confer drought Tolerance and Lead to Improved Corn Yields on Water-Limited Acres. *PNAS* **2007,** *104,* 16450–16455.

Nemchenko, A.; Kunze, S.; Feussner, I.; Kolomiets, M. Duplicate Maize 13-Lipoxygenase Genes are Differentially Regulated by Circadian Rhythm, Cold Stress, Wounding, Pathogen Infection, and Hormonal Treatments. *J. Exp. Bot.* **2006,** *57,* 3767–3779.

Pál, M.; Horváth, E.; Janda, T.; Páldi, E.; Szalai, G. Physiological Changes and Defense Mechanisms Induced by Cadmium Stress in Maize. *J. Plant Nutr. Soil Sci.* **2006,** *169,* 239–246.

Ribaut, J. M.; Jiang, C.; Gonzalez-de-Leon, D.; Edmeades, G. O.; Hoisington, D. A. Identification of Quantitative Trait Loci Under Drought Conditions in Tropical Maize. 2. Yield Components and Marker-Assisted Selection Strategies. *Theor. Appl. Genet.* **1997,** *94,* 887–896.

Saab, I. N.; Sachs, M. M. A Flooding-Induced Xyloglucan Endo-Transclycosylase Homolog in Maize is Responsive to Ethylene and Associated with Aerenchyma. *Plant Physiol.* **1996,** *112,* 385–391. DOI: https://doi.org/10.1104/pp.112.1.385.

Sánchez, B.; Rasmussen, A.; Porter, J. R. Temperatures and the Growth and Development of Maize and Rice: A Review. *Glob. Change Biol.* **2014,** *20,* 408–417.

Siebers, M. H.; Slattery, R. A.; Yendrek, C. R.; Locke, A. M.; Drag, D.; Ainsworth, E. A.; Bernacchi, C.; Ort, D. R. Simulated Heat Waves During Maize Reproductive Stages Alter Reproductive Growth but Have no Lasting Effect When Applied During Vegetative Stages. *Agric. Ecosyst. Environ.* **2017,** *240,* 162–170.

Voothuluru, P.; Anderson, J. C.; Sharp, R. E.; Peck, S. C. Plasma Membrane Proteomics in the Maize Primary Root Growth Zone: Novel Insights into Root Growth Adaptation to Water Stress. *Plant Cell Environ.* **2016,** *39,* 2043–2054.

Waananen, K. M.; Okos, M. R. Failure Properties of Yellow Dent Corn Kernels. *Trans. ASAE.* **1988,** *31,* 1816–1827.

Wilhelm, E. P.; Mullen, R. E.; Keeling, P. L.; Singletary, G.W. Heat Stress During Grain Filling in Maize: Effects on Kernel Growth and Metabolism. *Crop Sci.* **1999,** *39,* 1733–1741.

Zaidi, P. H.; Selvan, P. M.; Sultana, R.; Srivastava, A.; Singh, A. K.; Srinivasan, G.; Singh, R. P.; Singh, P. P. Association Between line *per se* and Hybrid Performance Under Excessive Soil Moisture Stress in Tropical Maize (*Zea mays* L.). *Field Crops Res.* **2007,** *101,* 117–126.

Zaidi, P. H.; Rafique, S.; Rai, P. K.; Singh, N. N.; Srinivasan, G. Tolerance to Excess Moisture in Maize (*Zea mays* L.): Susceptible Crop Stages and Identification of Tolerant Genotypes. *Field Crops Res.* **2004,** *90,* 189–202.

Zaidi, P. H.; Rafique, S.; Singh, N. N. Response of Maize (*Zea mays* L.) Genotypes to Excess Soil Moisture Stress: Morpho-Physiological Effects and Basis of Tolerance. *Eur. J. Agron.* **2003,** *19,* 383–399.

Zhang, Y.; Wang, L.; Liu, Y.; Zhang, Q.; Wei, Q.; Zhang, W. Nitric Oxide Enhances Salt Tolerance in Maize Seedlings Through Increasing Activities of Proton-Pump and Na^+/H^+ Antiport in the Tonoplast. *Planta* **2006,** *224,* 545–555.

CHAPTER 8

Biotic Stresses Affecting Crop Productivity

ABSTRACT

The biotic stresses such as insects and diseases affect the growth and productivity of maize. Several intensive research studies were undertaken in studying their effects. Recent research advances assisted in the identification of physiological, biochemical, and molecular mechanisms of these stress factors. The research advances that occurred are briefly reviewed in the following sections.

8.1 INSECT PESTS

Climate and weather have considerable effects on the development and distribution of insects. Therefore, the human-induced climatic change that arises from rising levels of atmospheric greenhouse gases could have a significant effect on agricultural insect pests. The present estimates of climate change indicated that global average annual temperatures would increase 1°C by 2025 and 3°C by the end of the next century. These changes in climate may cause variations in geographical distribution, improved overwintering, variations in population growth rates, increases in the number of generations, expansion of the development season, variations in crop-pest synchrony, changes in interspecific interactions, and increased risk of attack by migrant pests. In this paper, some of these effects were illustrated by studying the effect of climatic change on the maize European corn borer (*Ostrinia nubilalis*). By the Goddard Institute for Space Studies general circulation model, climatic changes were projected and European corn borer was estimated to occur up to 1220 km with an additional generation in nearly all regions where it was known to occur. Further, several priorities such as investigation of the effect of climatic variables on insect pests, long-term monitoring of pest population levels and insect behavior, deliberation of probable climatic fluctuations in research into pest management systems, and recognition of potential migrants were discussed (Porter et al., 1991).

The evolution of one of the major pests of maize *Dalbulus maidis* was reviewed. Previous studies reported that the *D. maidis* was originated during post-Colombian times, and its main host was teosintes (*Zea*) or the closely related gamagrasses (*Tripsacum*). The species of *Dalbulus* that specialize on gamagrasses were morphologically more primitive than *D. maidis* and *D. elimatus*. This indicated that *Tripsacum* is the ancestral host genus of *Dalbulus*. The evolution of *Dalbulus* might have driven by geographic isolation and host specialization. Afterward, when maize was first domesticated from its teosinte relatives, it adopted maize as a host. As the maize habitat is movable, the leafhoppers specializing maize should overwinter as hardy, mobile adults. Further, evidence suggested that maize pathogens such as *Spiroplasma kunkelii* and the maize bushy stunt mycoplasma have developed nonpathogenic, mutualistic associations with improved *Dalbulus* vector species (Nault, 1990).

8.2   DISEASE

In maize, Southern corn rust (SCR) caused by *Puccinia polysora* is a destructive disease. Inbred line Qi319 was found to be highly resistant to this disease. The inoculation test and genetic analysis of the Southern corn rust in Qi319 populations indicated that SCR resistance is regulated by a single dominant resistant gene, *RppQ*. Here, simple sequence repeat (SSR) analysis was conducted in an F_2 population obtained from the "Qi319×340" cross. Twenty SSR primer pairs were dispersed on chromosome10 and screened for resistant gene. Only two primer pairs, phi118 and phi 041, exhibited association with SCR resistance. On the basis of these results, eight novel SSR primer pairs flanking the region of primers, phi118 and phi 041, were picked chosen and verified for their association with *RppQ*. The findings showed that SSR markers, umc1,318 and umc2,018, were linked to *RppQ* with a genetic distance of 4.76 and 14.59 cM, respectively. On the other side of *RppQ*, away from SSR markers phi 041 and phi118, another SSR marker umc1,293 was linked to *RppQ* with 3.78 cM of genetic distance. As the five linkages SSR markers were detected on chromosome

10, the *RppQ* gene was expected to be situated on chromosome 10. Further, AFLP (amplified fragment length polymorphism) analysis was undertaken to fine map the *RppQ* gene. Out of 54 AFLP primer combinations analyzed, only one marker AF1 exhibited linkage with the *RppQ* gene in a genetic distance of 3.34 cM. Finally, the *RppQ* gene was located on the short arm of chromosome 10 between phi 041 SSR markers and AF1 AFLP marker (Chen et al., 2004).

Turcicum or northern corn leaf blight (NCLB) caused by *Exserohilum turcicum* is one of the major foliar diseases of maize. For this disease, varied sources of qualitative and quantitative resistance are available but qualitative resistance (*Ht* genes) is unstable. In the tropics, the qualitative resistance is overwhelmed by new virulent races or suffers from climatically sensitive expression. However, the expression of quantitative resistance is independent of the physical environment and has never succumbed to *Setosphaeria turcica* pathotypes in the field. This paper reviewed the consistency of the genomic positions of quantitative trait loci (QTL) regulating resistance to *S. turcica* across diverse maize populations, and the clustering of genes for resistance and other fungal pathogens or insect pests in the maize genome (Welz and Geiger, 2000).

Maize is susceptible to a wide range of diseases. To determine the dissemination of maize diseases such as northern leaf blight (NLB), common rust (CR), maize streak disease (MSD), gray leaf spot (GLS), head smut (HS), and common smut (CS) in Kenya, a survey was carried out in various agro-ecological zones (AEZ) of Kiambu, Embu, and Nakuru regions. The data were collected on the prevalence, incidence, and severity of these diseases. To characterize Maize streak virus (MSV) at the molecular level, maize leaf samples infected with MSD were collected. 100% prevalence of northern leaf blight was noticed in all regions. Higher incidences of NLB and GLS were found in high altitude areas, whereas CS and MSD were prevalent in the three regions. The sequences from the GenBank were compared with 797 nucleotides from the open reading frame (ORF) C2/C1 of MSV and results revealed the sequence correspondences of 99–100% with MSV-A strain. These findings unveiled the prevalence of major foliar diseases of maize in Kenya. Hence, suitable measures should be implemented to control these diseases and decrease related losses. The high percent sequence similarities of MSV indicated low variability that is good for breeders because developed resistant varieties can be adopted over a wider region (Charles et al., 2019).

Genetically engineered Bt maize hybrids with δ-endotoxin *cryIA* (b) were evaluated for the incidence and severity of *Fusarium* ear rot and the incidence of symptomless *Fusarium* infection in maize kernels. Treatments were manually infested with European corn borer (ECB) larvae and at specific growth stages of maize, insecticides were applied to limit ECB activity. In each hybrid, *Fusarium* symptoms and infection were found to be affected by the specific *cryIA(b)* that determined the tissue-specific expression of *cryIA*(b). When compared with near-isogenic hybrids lacking *cryIA(b)* genes, the incidence and severity of *Fusarium* ear rot symptomless kernel infection was less in hybrids expressing *cryIA(b)* in kernels. Further, *cryIA(b)* was found to reduce the infestation of ECB. The expression of *cryIA(b)* in plant tissues other than kernels did not consistently affect *Fusarium* symptoms or infection. Disease incidence exhibited a positive correlation with ECB damage to kernels. However, the application of insecticide also decreased *Fusarium* symptoms and infection when used to nontransgenic plants (Munkvold et al., 1997).

In maize, the Rp1 region was characterized as a complex locus that offers resistance to the fungus *Puccinia sorghi*, causing rust disease. Some alleles of Rp1 are meiotically unstable; however, the instability mechanism is not identified. In this study, the role of recombination in meiotic instability in maize lines homozygous for either Rp1-J or Rp1-G was investigated. Test cross progenies obtained from a line that was homozygous for Rp1-J and heterozygous at flanking markers were tested for susceptible progeny. Out of 9772 progenies, five progenies were found to be susceptible. All the five progenies had nonparental combinations of flanking markers. Three progenies showed one combination of recombinant flanking markers, whereas the other two had the opposite pair. Then, 5874 test cross progenies were subjected to an identical study with Rp1-G and 20 susceptible progenies were identified. From these progenies, nineteen were linked with flanking marker exchange, 11 and 8 of each recombinant marker combination. Therefore, the study findings suggested that unequal exchange is the main reason for meiotic instability in Rp1-J and Rp1-G (Sudupak et al., 1993).

Maize lethal necrosis (MLN) is a synergistic disease caused by maize chlorotic mottle virus (MCMV) and one of the viruses from potyviridae, for example, sugarcane mosaic virus, maize dwarf virus, Johnson grass mosaic virus, or wheat streak mosaic virus. The synergistic action of coinfecting viruses causes plant death or severe reduction in yield. The factors such as multiple maize crops per year, occurrence of maize thrips, and highly susceptible maize varieties were found to be related with MLN. It was found that the transmission of MCMV through seed and soil significantly increases the development and perpetuation of MLN epidemics. Multipronged approaches are required to prevent and control the MLN and additional studies are required to find and develop the best control measures (Redinbaugh and Stewart, 2018).

8.2.1 DISEASE RESISTANCE

Global productions of maize are affected largely by the occurrence of gray leaf spot disease caused by the pathogen *Cercospora zeae maydis* and *Cercospora zeina*. This pathogen exhibits a necrotrophic lifestyle, due to which no key genes were known largely. In a nested association analysis, genetic mapping has been done by Benson et al. (2015) to reveal the genetic architecture governing the resistance against this pathogen and to understand and develop the mechanisms conferring quantitative disease resistance loci. They detected 16 QTL for quantitative disease resistance, out of which 7 were new loci. These quantitative disease resistance loci exhibited allelic series in their effects that were either high or low in magnitude to that of a reference allele B73. Alleles identified at three QTL, qGLS1.04, qGLS2.09, and qGLS4.05, conferred disease reductions of larger than 10%. Interactions between loci were detected for three pairs of loci, including an interaction between iqGLS4.05 and qGLS7.03. QGLS1.04 was fine-mapped from an interval of 27.0 Mb to two intervals of 6.5 and 5.2 Mb, consistent with the assumption that multiple genes lie behind highly significant QTL identified by NAM. qGLS2.09, which was also associated with maturity (days to anthesis) and with resistance to southern leaf blight, was narrowed to a 4-Mb interval. The distance between major leaf veins was powerfully connected with resistance to GLS at qGLS4.05. NILs for qGLS1.04, when treated with the *C. zeae-maydis* toxin cercosporin, revealed that there was an increased expression of a putative flavin-monooxygenase (FMO) gene, a candidate detoxification-related gene underlying qGLS1.04.

In maize, multiple loci conferring small effects are linked with the quantitative resistance to plant pathogens. Poland et al. (2011) evaluated 5000 inbred lines of nested association mapping population to gain an insight into the quantitative resistance against northern leaf blight. They detected 29 QTLs, many of which possessed multiple alleles. Phenotypes varied in resistance against the leaf blight due to a buildup in a number of loci having minor additive effects. They have used 1.6 million single nucleotide polymorphisms (SNPs) for genome wide nested association mapping to detect candidate genes and were able to identify a number of multiple candidate genes, which were linked to confer plant defense. Some of these also included receptor-like kinase genes, which were identical to those involved in basal defense mechanisms. Their findings were in consistence with the assumption that in plants, the quantitative disease resistance is insured by a variety of mechanisms.

8.3   OTHER PESTS

In Rajasthan, maize monocropping, favorable soil type, ecological conditions, and unawareness of management practices encourage the infection of maize cyst nematode, *Heterodera zeae*. This nematode causes significant yield losses in maize. By considering its importance, the efficiency of botanicals such as *Calotropis procera* (latex), *Aloe vera* (gel), and *Euphorbia neriifolia* (latex) at 1, 2, and 4% w/v as seed treatment against maize cyst nematode was investigated. *Calotropis procera* at 4% was found to be more effective compared to *Calotropis procera* at 2% and *Aloe vera* at 4%. It was found that to decrease the infection of maize cyst nematode, *H. zeae*, the botanicals enhance the plant growth parameters of maize (Kumhar et al., 2018).

8.4   INTEGRATED PEST MANAGEMENT

Chicken egg white contains a glycoprotein avidin that reduces the vitamin biotin. In this study, it was reported that the presence of avidin in maize at ≥ 100 p.p.m. level is toxic and averts the development of storage pests by causing a biotin deficiency. This was verified by the prevention of toxicity with biotin supplementation. Therefore, this study suggested that avidin expression in food or feed grain crops could be applied as a biopesticide against many insects that damage stored produce (Kramer et al., 2000).

The corn rootworm (CRW; *Diabrotica* spp.) is a major pest of corn in the USA. Mostly, farmers follow crop rotation and use chemical insecticides against this insect. But these methods are ineffective because of resistance or behavioral modifications. Here, the transgenic maize developed by engineering a Cry3Bb1 gene from *Bacillus thuringiensis* (*Bt*) that control CRW was described. The Bt maize hybrids were found to be eight times more lethal to corn rootworm larvae than the wild-type protein. To develop Bt, maize-modified *cry3Bb1* gene was placed in a DNA vector under control of a root-enhanced promoter (4-AS1) and transferred into embryonic maize cells using microprojectile bombardment. Then, the performance of Bt maize was tested using molecular genetic characterization at protein expression levels and field performance (Vaughn et al., 2005).

Plants attacked by arthropod herbivores often release volatile compounds from their leaves that attract natural enemies of the herbivores. Similarly, maize roots release (E)-β-caryophyllene sesquiterpene that attracts entomopathogenic nematode against the *Diabrotica virgifera virgifera*, beetle larvae, which is currently invading Europe. Most of the European lines and wild maize ancestor, teosinte, release this sesquiterpene in response to the attack of *D. v. virgifera*, whereas North American maize lines do not. In a field experiment, a five-fold higher infestation of *D. v. virgifera* larvae was found on a maize variety that releases the sesquiterpene than on a variety that does not. However, the application of (*E*)-β-caryophyllene in the soil near the latter variety reduced the infestation of *D. v. virgifera* to less than half. Therefore, the efficiency of nematodes as biological control agents against root pests like *D. v. virgifera* could be enhanced by developing new varieties that release the attractant in sufficient amounts (Rasmann et al., 2005).

The harmful effect that the insect-resistant transgenic plants may cause on nontarget organisms, such as entomophagous arthropods (parasitoids and predators), is the major concern associated with the adoption of these plants. Though the regulatory bodies need data regarding the possible risk of releasing transgenic plants in the environment, up to now, specific protocols have not been proposed for evaluating the risks of insect-resistant transgenic crops on entomophagous insects. In this paper, the procedure to be followed for evaluating the effects of insect-resistant plants on entomophagous insects was proposed with Bt maize that codes for the Cry1Ab toxin. First, the entomophagous insects that play a key role in regulating maize pests, and which may be at risk because of transgenic plants should be determined. As the risk which transgenic plants cause to entomophagous insects depends on both, their contact, and their sensitivity to the insecticidal protein, at what level organisms are subjected to the transgene compound should be determined at the second step. Exposure is related with the feeding behavior of phytophagous and entomophagous insects along with the tissue- and cell-specific temporal and spatial expression of the insecticidal protein. So, to assess toxicity the sensitivity tests should be performed for the organisms that could be exposed to the insecticidal protein (Dutton et al., 2003).

Augmentative biological control is a method of insect control where natural enemies are mass-reared in biofactories and released periodically in large numbers to control the pests immediately. This method has been successful, ecologically and economically, and observed as a perfect substitute for chemical pest control in crops such as maize, cotton, sugarcane, soybean as well as in fruit orchards, vineyards, and greenhouses. During the past decades, it has transferred from a cottage industry to professional production. Several effective natural enemies' species have been discovered and 230 are commercially available currently. Further, the industry established quality control guidelines, mass production, shipment, and release methods along with suitable guidance for farmers. But this method is useful on a small acreage. So, in this paper, the causes for the limited application of augmentative biological control in addition to the trends in research and its application were studied (van Lenteren, 2012).

8.5   RESEARCH ADVANCE IN BIOTIC STRESS MANAGEMENT

Nested association mapping (NAM) has the ability to resolve complex, quantitative traits on their underlying loci. In this study, the maize NAM population containing 5000 recombinant inbred lines (RILs) from 25 families were examined for resistance against southern leaf blight (SLB) disease. Joint-linkage analysis detected 32 QTLs with small, additive effects on SLB resistance. Then for maize HapMap SNPs, genome-wide association analysis was performed by assigning founder SNP genotypes onto the NAM RILs. Many of these SNPs were found within or nearby the sequences homologous to genes, which were earlier appeared to be concerned with plant disease resistance. SNPs both within and outside of QTL intervals were related with variation for SLB resistance. Partial linkage disequilibrium was noted around some SNPs linked with SLB resistance. These findings indicated that the maize NAM population facilitates high-resolution mapping of genome regions (Kump et al., 2011).

Many plant disease resistance genes that have been isolated were encoding proteins with a putative nucleotide binding site and leucine-rich repeats (NBS-LRR resistance genes). On the basis of conserved motifs in and around the NBS of known NBS-LRR resistance oligonucleotide primers were designed. Then, these primers were utilized to amplify sequences from maize genomic DNA by means of polymerase chain reaction (PCR). Eleven classes of noncross-hybridizing sequences were found. These sequences had expected products with high levels of amino acid resemblance to NBS-LRR resistance proteins. Further, 20 restriction fragment length polymorphism (RFLP) loci were mapped in maize using these maize resistance gene analogs (RGAs) and one RGA clone acquired earlier from wheat as probes. Few RFLPs mapped the genomic regions with virus and fungus resistance genes and the RGA loci exhibited a complete co-segregation with the rust resistance loci *rp1* and *rp3*. These findings suggested that the RGA clones may hybridize to resistance genes (Collins et al., 1998).

8.6 CONCLUSIONS

Several biotic stresses such as insects and diseases have their effect on plants. The research advances that have been undertaken for understanding the effects of these stress factors were discussed in this chapter. Biotic stresses caused by insects and diseases are being intensified because of favorable conditions. To overwhelm this problem and meet the worldwide demand for food, tolerant crops must be developed by means of suitable technologies. Physiological aspects, mechanisms of tolerance, and management strategies for better crop production must be comprehensively investigated to effactually use such technologies.

KEYWORDS

- maize
- biotic stress
- research advances
- insects
- diseases

REFERENCES

Bahadori, A., Mobasser, H. R.; Ganjali, H. R. Influence of Water Stress and Plant Density on some Characteristics in Corn. *Biological Forum – Intern. J.* **2015,** *7*(1), 673–678.

Benson, J. M.; Poland, J. A.; Benson, B. M.; Stromberg, E. L.; Nelson, R. J. Resistance to Gray Leaf Spot of Maize: Genetic Architecture and Mechanisms Elucidated through Nested Association Mapping and Near-Isogenic Line Analysis. *PLOS Genet.* **2015,** *11*, e1005045.

Chen, C. X.; Wang, Z. L.; Yang, D. E.; Ye, C. J.; Zhao, Y. B.; Jin, D. M.; Weng, M. L.; Wang, W. B. Molecular Tagging and Genetic Mapping of the Disease Resistance Gene *RppQ* to Southern Corn Rust. *Theor. Appl. Genet.* **2004,** *108*, 945–950.

Collins, C.; Webb, C. A.; Seah, S.; Ellis, J. G.; Hulbert, S. H.; Pryor, A. Isolation and Mapping of Disease Resistance Gene Analogs in Maize. *Mol. Plant Microbe Interact.* **1998,** *11*, 968–978.

Charles, A. K.; Muiru, W. M.; Miano, D. W.; Kimenju, J. W. Distribution of Common Maize Diseases and Molecular Characterization of Maize Streak Virus in Kenya. *J. Agric. Sci.* **2019,** *11*, 47–59.

Dutton, A.; Romeis, J.; Bigler, F. Assessing the Risks of Insect Resistant Transgenic Plants on Entomophagous Arthropods Bt-Maize Expressing Cry1Ab as a Case Study. *BioControl* **2003,** *48*, 611–636.

Kramer, K. J.; Morgan, T. D.; Throne, J. E.; Dowell, F. E.; Bailey, M.; Howard, J. A. Transgenic Avidin Maize is Resistant to Storage Insect Pests. *Nat. Biotechnol.* **2000,** *18*, 670–674.

Kump, K. L.; Bradbury, P. J.; Wisser, R. J.; Buckler, E. S.; Belcher, A. R.; Oropezas-Rosas, M. A.; Zwonitzer, J. C.; Kresovich, S.; McMullen, M. D.; Ware, D.; Balint-Kurti, P. J.; Holland, J. B. Genome-Wide Association Study of Quantitative Resistance to Southern Leaf Blight in the Maize Nested Association Mapping Population. *Nat. Genet.* **2011,** *43*, 163–168.

Kumhar, R. N.; Baheti, B. L.; Chandrawat, B. S. Eco-Friendly Management of Maize Cyst Nematode, Heteroderazeae on Maize by Use of Botanicals. *Int. J. Curr. Microbiol. Appl. Sci.* **2018,** *7*, 199–204.

Munkvold, G. P.; Hellmich, R. L.; Showers, W. B. Reduced *Fusarium* Ear Rot and Symptomless Infection in Kernels of Maize Genetically Engineered for European Corn Borer Resistance. *Phytopathology* **1997,** *87*, 1071–1077.

Nault, L. R. Evolution of An Insect Pest: Maize and the Corn Leafhopper, A Case Study. *Maydica* **1990,** *35*, 165–175.

Poland, J. A.; Bradbury, P. J.; Buckler, E. S.; Nelson, R. J. Genome-Wide Nested Association Mapping of Quantitative Resistance to Northern Leaf Blight in Maize. *Proc. Natl. Acad. Sci. U.S.A.* **2011,** *108*(17), 6893–6898.

Porter, J. H.; Parry, M. L.; Carter, T. R. The Potential Effects of Climatic Change on Agricultural Insect Pests. *Agric. Forest Meteorol.* **1991,** *57*, 221–240.

Rasmann, S.; Köllner, T. G.; Degenhardt, J.; Hiltpold, I.; Toepfer, S.; Kuhlmann, U.; Gershenzon, J.; Turlings, T. C. J. Recruitment of Entomopathogenic Nematodes by Insect-Damaged Maize Roots. *Nature* **2005,** *434,* 732–737.

Redinbaugh, M. G.; Stewart, L. R. Maize Lethal Necrosis: An Emerging, Synergistic Viral Disease. *Ann. Rev. Virol.* **2018,** *5,* 301–322.

Sudupak, M. A.; Bennetzen, J. L.; Hulbert, S. H. Unequal Exchange and Meiotic Instability of Disease-Resistance Genes in the Rp1 Region of Maize. *Genetics* **1993,** *133,* 119–125.

van Lenteren, J. C. The State of Commercial Augmentative Biological Control: Plenty of Natural Enemies, But a Frustrating Lack of Uptake. *BioControl* **2012,** *57,* 1–20.

Vaughn, T.; Cavato, T.; Brar, G.; Coombe, T.; DeGooyer, T.; Ford, S.; Groth, M.; Howe, A.; Johnson, S.; Kolacz, K.; Pilcher, C.; Purcell, J.; Romano, C.; English, L.; Pershing, J. A Method of Controlling Corn Rootworm Feeding Using a *Bacillus thuringiensis* Protein Expressed in Transgenic Maize. *Crop Sci.* **2005,** *45,* 931–938.

Welz, H. G.; Geiger, H. H. Genes for Resistance to Northern Corn Leaf Blight in Diverse Maize Populations. *Plant Breed.* **2000,** *119,* 1–14.

CHAPTER 9

Methods of Cultivation

ABSTRACT

This chapter deals with the different methods of maize crop cultivation. Maize cultivation includes various operations from land preparation to marketing which are briefly discussed here. Besides this, the research advances made in maize cultivation are discussed.

9.1 SELECTION OF CULTIVARS

An experiment was conducted to evaluate corn cultivars for their adaptability and stability under seven diverse environments in the Amazonas state. Using REML/Blup methodology, the genetic parameters of cultivars were assessed. On the basis of expected genetic value and harmonic mean of the genetic values relative performance, cultivars with high adaptability and stability were selected. In spite of the presence of genotype × environment interaction, corn cultivars with high adaptability and stability were found. The single-cross hybrid BRS 1055 and the synthetic varieties Sint 10771, Sint 10781, and Sint 10699 had high productivity and high stability in this state. The hybrid BRS Caimbé exhibited specific adaptability to cropping in upland environments of the Amazonas state, Brazil (de Oliveira et al., 2017).

9.2 LAND PREPARATION

Under irrigated conditions in the arid regions of southern Africa, the compaction of subsoil and plowed layer on deep fine sandy soils caused by the conventional tillage practices restricts root growth. During seedbed preparation, this plowed layer is recompacted by the uncontrolled wheel traffic on the surface. In this study, the conventional tillage practice with two controlled wheel-traffic practices, each containing deep ripping of the subsoil followed by the control of wheel traffic to specific lanes were compared during seedbed preparation. To create favorable conditions for root growth the traffic lanes were loosened in one treatment, while in the other they kept intact. Significant increase in root depth, root density in the subsoil, water-use efficiency, and a 30% yield increase was observed in deep ripping and controlled traffic plots. The positive yield effect of ripping was found to be decreased by about 4% in the loosened traffic lanes (Bennie and Botha, 1986).

Conservation tillage is gaining more importance among farmers because of its lower costs of production than conventional tillage. Here, the long-term effects of subsoiling tillage (ST), no tillage (NT), and conventional tillage (CT) on soil properties and crop yields were studied over a period of eight years (2000–2007). The water stability of macro-aggregates at 0–0.30 m soil depth was greater for ST (22.1%) and NT (12.0%) than for CT. Significantly improved aeration porosity was observed in ST (14.5%) and NT (10.6%). After eight years, 0.8–1.5% lower soil bulk density was recorded in ST and NT treatments than in CT. As a result, higher crop yields were obtained in ST and NT plots than in CT plots because of the improved physical and chemical properties of soil. Among the conservation tillage treatments, though the economic benefits of ST and NT were similar, the ST had better effects on crop yields than NT. The crop yield in ST plots was 1.5% higher than NT. Their study deliberated that ST is the most appropriate conservation tillage practice for regions in the Beijing area (Zhang et al., 2009).

A field trial was undertaken to evaluate the maize hybrids' performance under different planting methods and nitrogen levels. The split plot experiment was designed by applying four planting techniques, namely, raised bed system, flat sowing with zero tillage (FSZT), flat sowing with minimum tillage, and flat sowing with conventional tillage in main plots and combination of two maize hybrids, namely, HQPM-1 and HQPM-5 and three nitrogen levels, namely, 120, 150, and 180 kg N ha^{-1} in subplots. Maize planted with FSZT exhibited the maximum nutrient uptake. Similarly, the maize hybrid HQPM-5 showed significantly higher nitrogen uptake when compared to HQPM-1 hybrid. Among the nitrogen levels, N uptake was found to improve with increasing levels of nitrogen from 120 kg N ha^{-1} to 180 kg N ha^{-1}. Similar effects of different treatments were recorded on phosphorus and potassium uptake by maize. The nitrogen, phosphorus, and potassium (NPK) uptake was found to increase significantly with nitrogen application (Tiwari et al., 2018).

9.3   TYPES OF SOWING

To study the influence of interrow management on root growth and rooting pattern in maize, an experiment was conducted. About 5-cm deep postplanting cultivation, 2-cm thick straw mulch, interrow compaction, and control were applied as treatments on sandy loam and loamy sand soils. It was found that straw mulch and cultivation improved root growth in the upper 15 cm of soil and increased the lateral spread of roots. Roots below 15 cm were less in mulch treatment than the control. In the surface layers, interrow compaction inhibited the lateral spread of roots leading to downward growth of roots. Reduction of soil temperature by 2.6°C was observed at the 5-cm soil depth in mulched plots. Further, the reductions in soil temperature and moisture losses in mulched plots were found to be associated with greater lateral spread and improved root growth in the surface layers resulting in improved plant growth and higher yield (Chaudhary and Prihar, 1974).

9.3.1   EFFECT OF SOWING DATES ON SOURCE–SINK RELATIONSHIP

In many crops and maize, there were decreases in yield with a delay in the sowing dates. The declined yields may be the resultant of reduced seed number, size, and unfilled grains. In addition, a disruption in the supply of assimilates to the developing kernels, or in the activity of sinks might also contribute to declining yields. Limitation of either the source or the sink may contribute to the reduced yields also. Though a number of studies have been performed on the effects of delayed sowing on yields, yet it remained unclear, whether, the declined yields are the resultant of a decline in the source strength and activity or due to a decline in sick strength and activity. Optimization of crop management practices is possible, only when there is a clear understanding of the relationships of source–sink and also when critical processes are identified. Effective breeding strategies can be developed. Bonelli et al. (2016) assessed the effect of delayed sowing date on the grain yield components as well as on source–sink relationships at Balcarce, Argentina, for three consecutive cropping season years. The crop was sown at varied sowing dates from October to January, during which there was a large seasonal photothermal variation. Late sowing in January resulted in a drastic decline in grain yields. The proportionate decrease in grain number was less in comparison to a decline of grain weight per grain, with a delay in sowing date. Large differences in grain yield are found to be linked closely to that of the harvest index. With a delay in the sowing dates, there was a decline in the accumulation of dry matter (DM) in the postsilking period, which resulted in variations in the harvest index. With variation from early to late sowing dates, there were differences in the source capacity and sink strength. The differences in source capacity were more pronounced rather than the variations in the sink strength, particularly during the grain-filling period, there was much reduction in the source/sink ratio. Findings specify that crop growth during the grain-filling period is limited to a large extent by a limitation in sink strength at the early sowing days, while the limitation of photosynthetic source capacity acted as a limiting factor during the time of late sowing.

9.3.2   EFFECT OF SOWING DATE ON YIELD

Tsimba et al. (2013) evaluated the effects of varied planting dates (18 September to 15 December) in six maize hybrids of varying maturity levels in Waikato and Manawatu regions of New Zealand (2006 and 2007) on leaf growth rates, grain yields, and yield components. An increase in the daily mean temperature (13–19°C) at once preceding to tassel initiation resulted in a decline in the leaf number by 0.1 leaf °C^{-1} was observed. However, these hybrids exhibited the largest leaf area indices at daily temperatures of 17–19°C. Early plantings of maize hybrids resulted in maximum grain yields in the Waikato regions, possessing a lower latitude environment. In contrast, declined yields, resulted from the early plantings at Manawatu regions were due to the prevalence of low temperatures during the spring season, due to these low temperatures, the hybrids grown under early plantings produced canopies of smaller size. Though early hybrids produced consistent yields across varied planting dates, late hybrids when planted early out yielded these early hybrids. The late hybrids when planted early had a better balanced source-sink ratio. Further, late plantings also resulted in declined yields due to the existence of low temperatures (15–18°C) and radiation (11–20 MJ m^{-2} day^{-1}) during the grain-filling periods. Grain yield was found to be in strong association with the kernel number and weight in these hybrids. Response of these hybrids to planting dates, water stress, temperature, and irradiance was more pronounced in late hybrids in their kernel number, kernel weight as well as grain yield. Water stress effects seen on the grain filling was found to have minimal effects during the late plantings. Kernel number was more affected by water stress, though a stable performance was found among these hybrids in their kernel weight. In conclusion, the results revealed that late plantings in maize result in a decline in the biomass production and harvest index.

9.4 WATER MANAGEMENT

To focus on research and make strategies for food security, the quantification of the available gap between average farmer yields and yield potential (Y_p) is important. In this study, the yield potential, yield gaps, and the effect of agronomic practices on both parameters were quantified in irrigated maize systems of central Nebraska. Three years data with field-specific values for yield, applied irrigation, and N fertilizer rate were used for the analysis. By means of a maize simulation model together with actual and interpolated weather records yield potential was estimated. Then from a subset of fields, complete data on crop management was collected. The variance between actual yields and simulated Y_p for each field-year observation was taken into account to estimate yield gaps. Further, long-term simulation analysis was carried out to assess the sensitivity of Y_p to variations in selected management practices. Findings revealed that present irrigated maize systems reached the Y_p ceiling. In the USA, the mean actual yield varied from 12.5 to 13.6 Mg ha^{-1} across years and the mean N fertilizer efficiency was 23% greater than average efficiency. It was found that rotation, tillage system, sowing date, and plant population density were the most sensitive factors and had a direct correlation with actual yields. Time trends in mean farm yields from 1970 to 2008 showed that yields have not improved during the past eight years. Based on present crop management practices the average yield reached ~80% of Y_p ceiling during this period. Simulation analysis has shown that Y_p can be improved by higher plant population densities and by hybrids with longer maturity, but the implementation of these practices may have some constraints like trouble in planting and harvest operations because of wet weather and snow, additional costs of seed and grain drying, and high frost and lodging risks. Finally, the study suggested that the maize produced under irrigated conditions reached the Y_p ceiling and high levels of N-use efficiency and further improvement can be made by fine-tuning present management practices (Grassini et al., 2011).

A study was carried out to investigate the effect of different irrigation levels on maize water productivity in semiarid tropical climate. Treatments used were surface irrigation at 0.6 IW/CPE ratio (T1), 0.8 IW/CPE ratio (T2), 1.0 IW/CPE ratio (T3), 1.2 IW/CPE ratio (T4), drip irrigation at 0.6 Epan (T5), drip irrigation at 0.8 Epan (T6), drip irrigation at 1.0 Epan (T7), and drip irrigation at 1.2 Epan (T8). The findings unveiled that drip irrigation 0.6 Epan has provided the highest water productivity (1.34 kg m^{-3}) with consumption of 3130 m^3 of water. The lowest water productivity was obtained in surface irrigation scheduled at 1.0 (0.84 kg m^{-3}) and 1.2 (0.84 kg m^{-3}) IW/CPE ratio with a water consumption of 4670 and 5170 m^3 of water, respectively (Roja et al., 2017).

9.5 FERTILIZER MANAGEMENT

The maize cropping pattern can be diversified by cultivating baby corn. In this study, the baby corn variety (VL-78) performance which is suggested for the north hill zone was assessed under different fertility levels. The highest cob yield (without husk) of 20.60 q ha^{-1} along with the maximum number of cobs plot^{-1} (326) were recorded in plots applied with farmyard manure (FYM) at 6 t ha^{-1} plus 150% recommended dose of fertilizer (RDF) (225N:90P$_2$O$_5$:60K$_2$O kg ha^{-1}), but the FYM application at 6 t ha^{-1} plus state recommended dose of nitrogen:phosphorus:potassium (N:P:K) at 90:60:40 kg ha^{-1} was found to be statistically on par with the best treatment and provided a cob yield of 19.85 q ha^{-1}. Further nutrients application did not show any notable enhancement in morphological characters. The increased cob length (10.90 cm) was observed in plots applied with only 150% of RDF, whereas maximum cob girth without husk (18.30 mm) was found in plots applied with 125% of RDF. Under temperate conditions, baby corn variety VL-78 cultivation with an application of N:P:K at 90N:60P:40K, kg ha^{-1} combined with 6 t ha^{-1} FYM unveiled a highest B:C ratio of 1:1.59 and hence specified as a best cultivation practice (Lone et al., 2013).

The long-term effect of organic and inorganic fertilizers on soil health and grain quality was investigated by monitoring the enzyme activities and chemical properties of soil. Eight treatments, namely, a control; nitrogen and phosphorus (NP), nitrogen and potassium (NK), phosphorous and potassium (PK) and NPK; FYM alone, and addition of FYM at two different doses to NPK (NPK + FYM and 1/2 NPK + FYM) were applied. The increased dehydrogenase activity of soils and highest yield were recorded in plots applied with NPK + FYM and 1/2 NPK + FYM. The mineral fertilized plots had the highest grain protein content and test weight. The increased enzymatic activities except urease were recorded in plots applied with manure, while the urease activity was high in mineral N applied plots. The higher activities of urease, phosphatase and diesterase activities were observed in 1/2 NPK + FYM treated plots. Therefore, the long-term application of inorganic nutrients together with FYM was found to improve grain mineral composition and yield (Saha et al., 2008).

Nutrient management effect on nutrient uptake and economics of maize was investigated under different tillage practices. The main plot comprised of three tillage practices, namely, zero tillage, conventional tillage (CT), and bed planting (BP), and four different level of nutrient management, namely, RDF (120, 60, and 50 kg ha^{-1} N, P$_2$O$_5$, and K$_2$O), site-specific nutrient management (SSNM) on the basis of nutrient expert and farmers practice (FP) (150% of RDF + 10 t FYM) were applied in subplots. The highest gross returns (106,396 Rs ha^{-1}), net returns (64,111 Rs ha^{-1}), and total available nutrient uptake (N, P, K Fe, and Zn) were noted in bed-planting plots. Significantly higher net returns (63,523 Rs ha^{-1}), B:C ratio (1.89), and total available nutrient uptake (N, P, K, Fe, and Zn) were noted in SSNM when compared with other nutrient management practices (Kumar et al., 2019).

9.6 PHYSIOLOGICAL MATURITY

Farmers use various methods to estimate the maturity status of maize grains. In this study, the utility of the kernel milk line as a visual means of observing grain maturity in maize was evaluated. In five maize hybrids, the movement of kernel milk line was observed and compared with the moisture loss, accumulation of dry weight, and development of black layer in the kernels. To characterize the movement of kernel milk-line under prematurity stress, leaves from four hybrids were removed at different kernel development stages. Under normal conditions, kernel milk line disappearance was coincident with the development of the black layer and physiological maturity (PM). Though it was easy to ascertain the day on which milk in kernels was no longer present than to find when the placental region turns black. Therefore, the movement of the milk line could be used in monitoring the maturity process (Fig. 9.1). At the half-milk stage, kernels attained 90% of their dry weight, with 40% moisture, while the completely defoliated plants accumulated 95% of normal dry weight. At this stage, plants were 2–3 weeks from the optimum date to initiate harvest. So the study recommended that both the black layer development and kernel milk line disappearance can be used as indicators of maturity in maize. However, the kernel milk line was more useful in timing grain maturity before PM and the development of the black layer was the more dependable indicator of PM (Afuakwa and Crookston, 1984).

Previous studies have precluded the usage of grain moisture (GM) as an estimate of PM in maize. The cause of this inconstancy in GM values is still uncertain. Many studies suggested that the source–sink ratio may have an influence on the dynamics of kernel water relations during grain filling and, thus, GM at PM. To examine this probability, the reproductive sink capacity or the assimilate availability during grain filling was manipulated in maize treatments. Throughout the grain-filling kernel dry weight, water content, and the dry weight to water content (D–W) ratio, were observed. In each treatment gain moisture at PM was estimated by using a bilinear model relating dry weight and GM. It was found that the restriction of source capacity during grain filling increases GM at PM. However, the increment of source capacity per kernel did not show any effect on GM at PM. A single model explained the association between dry weight and GM for all hybrids and treatments without reduction of source capacity during grain filling. This model assessed 34.9% of GM at PM. These findings suggested that the measurement of GM at the late grain-filling stage could be a reliable estimate of PM when the source capacity was not severely limited. All the source–sink treatments had a similar D–W ratio of developing kernels till PM. However, premature termination of grain filling caused by defoliation resulted in desiccation of kernels and thus, increased the D–W ratio (Sala et al., 2007).

An experiment was carried out to determine the effect of harvesting maize at various stages of grain maturity on the yield and quality of maize grain and stover. The maize harvested at 28–30, 20–23, and 10–12% of GM content was termed as stages I, II, and III, respectively. At GM content of 12.5% grain yield exhibited an increasing trend. However, cob, stover, total crop residue, and total biomass DM yield exhibited a decreasing trend with increasing maturity stage and this decline in stover yield was due to leaf loss. With the reduction in GM content from 30% to 10%, a significant decrease in crop residue–grain ratio and leaf–stem ratio and a significant increase in the harvest index and grain hectoliter weight were observed. The highest stover ash content was obtained in maize harvested at stage I. Maize harvested at stage III had significantly lower crude protein and higher neutral detergent fiber and cellulose contents than those at stages I and II. In stage I, washing loss, potential degradability, and effective DM degradability at 0.03 h^{-1} rumen outflow rate were higher than in stages II and III. Further, in stage I higher gas production was recorded after 3, 6, 12, 24, 48, and 72 h of incubation. The reduction in crude protein and an increase in fiber concentration with increasing maturity stage contributed to a decrease in the stover-nutritive value. This in-turn reflects the variations in stover morphological composition and losses of nutrients within the morphological fractions with increasing maturity stage (Tolera et al., 1998).

Early-maturing maize genotypes yield 15–30% less than the late-maturing genotypes. A strategy for improving the yields of early-maturing groups involves the assessment and improvement of grain-filling traits, acting as secondary traits of selection for higher yields. Gasura et al. (2013) investigated 44 hybrids of tropical maize, in Zimbabwe, to increase the grain yields of early maturing maize, keeping in view the possibility of using the rate of grain filling as well as duration of grain filling under irrigated and nonirrigated environments. Most of the hybrids were above the trend line, though there was variation in the grain-filling rates and effective grain-filling duration, which were found to be negatively correlated. The earliest-maturing hybrid took 127 days to reach PM and produced grain yields comparable to those of the medium-maturing genotypes (7 t ha^{-1}). It had a high rate of 2.40 g plant^{-1} day^{-1} (18% higher than those of the low-yielding hybrids) of grain-filling rate and considerably a larger and effective grain-filling duration. The grain-filling rate and effective filling duration had high coefficients of genetic determination and exhibited positive correlations with that of grain yield. The study has shown that it is possible to develop high-yielding early-to-medium-maturing hybrids of maize on the basis of favorable combining ability values for rate of grain filling and grain durations.

9.6.1   PATTERN OF LEAF SENESCENCE IN MAIZE HYBRIDS

Improvements in grain yields of maize were related to the delayed rate of leaf senescence in various hybrids. Valentinuz and Tollenaar (2004) quantified the vertical profile of leaf senescence during the stage of grain-filling period in maize hybrids, Pride 5, Pioneer 3902, and Pioneer 3893, at Elora, Canada. Pioneer 3902 and Pioneer 3893 recorded maximum leaf area index (LAI) values at the silking stage. It was found that these hybrids exhibited a linear rate of leaf senescence of 0.44% day^{-1} during the grain-filling period; however, the rate of leaf senescence was only 1.87% day^{-1} during both first and second half of the grain-filling period. In Pride 5, an older hybrid of maize a threefold higher rate of leaf senescence was recorded during this period. In the initial first half of the grain-filling period, there was an increase in the leaf senescence rate with an increase in the plant population density. In contrast to this, in the second half of the grain-filling period, the rates of leaf senescence exhibited a decline with an increase in the plant population density in the Pride 5 hybrid. Despite these leaf senescence rates that were evident in Pioneer 3902 and 3893, in the medium population densities of these hybrids, there were lower rates of leaf senescence. The pattern of leaf senescence observed in the second half of the grain-filling stage was a top-to-bottom pattern of leaf senescence, wherein those leaves which were present at the central section of the canopy were found to be the last ones to exhibit senescence. This phenomenon was more clearly evident in Pioneer 3902 and Pioneer 3893, new maize hybrids.

9.7   METHOD OF HARVESTING

An experiment was undertaken to ascertain the ideal harvest time (IHT) for the seed production of XY335 and ZD958 maize cultivars (Fig. 9.1). In eight diverse environments, six seed-associated traits were assessed in seeds harvested at 11 different harvest stages. The vigor traits associated with seed were standard germination (SG), accelerated aging germination (AAG), and cold test germination (CTG), physiological traits were hundred seed weight (HSW) and seed moisture content (SMC) and an ecological trait associated with seed was above 10°C accumulated temperature from pollination to harvest (AT10ph). The harvest stage exhibited a significant effect on all the traits. The SG, AAG, CTG, and HSW responses to postpone harvest stage-fit quadratic models, whereas SMC and AT10ph fit linear models. The IHT for XY335 was 57.97 DAP, and for ZD958, it was 56.80 DAP. At IHT, SMC and AT10ph were 33.15% and 1234°C for XY335 and 34.98% and 1226°C for ZD958. The period to attain the maximum IHT was 5 days earlier than the HSW. AT10ph showed a strong association with the seed vigor traits. Along with the fact that AT10ph was less influenced by the environment, these findings put forward that AT10ph may be a novel indicator for ascertaining the IHT (Gu et al., 2017).

FIGURE 9.1   Maize cob ready to harvest.

Different methods of maize harvesting and threshing were evaluated in the Hoshiarpur district of Punjab. The harvesting and threshing methods such as manual harvesting and threshing with conventional maize thresher (T1); manual harvesting and threshing with maize dehusker-cum thresher (T2); and harvesting, threshing, and cleaning using self-propelled maize combine harvester (T3) were used as treatments for the study. Threshing efficiency for conventional thresher ranged from 97% to 99%, for maize dehusker-cum-sheller from 97% to 98%, and for self-propelled maize combine harvester from 95% to 97%. Maize harvested with combine harvester had the highest total grain losses (2–4%) and conventional maize thresher had the least (0.5–1.5%) grain loss. The net cost of harvesting per hectare was highest (Rs. 9000) in T3 and lowest (Rs. 2650) in T1. Manual harvesting and threshing with conventional maize thresher provided the highest total income from the maize residue per hectare (Rs. 5250). However, there was a saving of 100–140 man h ha^{-1} as labor requirement for dehusking of the crop in T2 and T3. Further, the maize residue left in harvesting, threshing, and cleaning using self-propelled maize combine harvester can be incorporated into the soil using a rotovapor which improves the soil health (Singh, 2014).

9.8 DRYING-MANAGEMENT OF GRAIN MOISTURE CONTENT

Maize grains from different IMC were dried in a domestic microwave oven and studied for their dry milling characteristics. The grains were also dried in a convective dryer at 65–90°C. The drying rate curve exhibited a characteristic case of moisture loss from grains by diffusion. Then, the dried samples were ground in a hammer mill and the Bond's work index was analyzed. The decrease in Bond's work index was observed with the increment in microwave drying duration. To measure the viscosity 10% suspensions were prepared by heating flour in water at 80 and 90°C and then cooled. It was observed that viscosity decreases with the increment in microwave drying time of the grains. The color analysis has shown that microwave-dried flour samples were brighter than the control and convective dried samples. The grains and the ground products had comparable protein and starch content (Velu et al., 2006).

A study was conducted to ascertain the time of net loss of water from the kernels and ear initiation and how the ear moisture-content (% wet-weight basis) can be used as a scale for kernel development during grain-filling. The associations between ear moisture, PM, and the environment were also measured. Maize cultivar Pioneer 3901 was observed from silking to harvest maturity and DM and water content of kernels, rachii, and ears were measured. It was found that water loss occurs in two phases. The first phase which ends at PM, exhibits a steady rate of water loss and is elucidated as "developmental" loss of water related with grain filling.

The second phase, initiating at PM, had a decreasing rate of water loss indicating a drying process. A similar water loss pattern was observed in 17 genotypes, with maximum ear moisture content of 80% and 60% in all the cases. With decreasing ear moisture, an increment in dry-weight per kernel was noted, this enables a direct and accurate estimation of ear moisture-content at the commencement of PM. This analysis method was showed to be valuable for assessing the effect of the environment on grain-filling in a single cultivar. The association between dry-weight per kernel and ear moisture was found to be constant for Pioneer 3901 studied over five seasons, but it exhibited variations at the grain-filling stage. Therefore, the breeding strategies aimed at reducing the GM at harvest should consider the absolute levels of water in the ear and the more usual expression of moisture content on a wet-weight basis (Brooking, 1990).

Moisture content, incidence of mold, ergosterol, and level of myco-toxin contamination were measured in 197 samples of maize collected from different agroclimatic regions of Karnataka. The moisture content was ascertained by using the hot-air oven method. In 34 maize samples, the moisture content (15–18%) above the permissible limit for safe storage was recorded. By using HPLC ergosterol was quantified and it was found in many samples gathered from rural areas of Karnataka regardless of the moisture content. Mold enumeration based on blotter and agar-plating methods unveiled the existence of 24 diverse species of both field and storage molds in samples. Mycotoxin analyses performed by means of monoclonal antibody-based enzyme-linked immunosorbent assay (ELISA) and thin-layer chromatography unveiled the mycotoxin contamination in 69 (34.8%) samples. The presence of molds and high levels of mycotoxins in many samples indicated the requirement of accurate surveillance and monitoring in maize (Janardhana et al., 1999).

9.9 PACKING AND STORAGE

Maize is a source of livelihood for millions of people in Africa. Poor manage-ment of harvested produce that causes 14% to 36% loss of maize grains is one of the main constraints for the improvement of food and nutritional security in Africa. The removal of part of the supply from the market due to postharvest losses results in high food prices. Therefore, the reduction of postharvest losses, in maize is crucial to increase food production without increasing the load on the natural ecosystem. To solve the problem of postharvest losses the effective cooperation and linkage among the research, extension, agro-industry, marketing system, and favorable policy environment is required (Tefera, 2012).

Prostephanus truncatus, Sitophilus zeamais, and *Rhyzopertha dominica* are major storage pests of maize that cause about 30% of grain losses. For postharvest storage of cowpea, a simple low-cost triple bagging technology called purdue improved crop storage (PICS) has been developed. Here, the applicability of PICS to maize storage was evaluated. The naturally infested maize with high levels of *P. truncatus, S. zeamais,* and *R. dominica* infestation were stored in PICS bags at seven locations. After 6.5 months of storage insect mortality of 95–100% was observed at all sites. Seed viability and germination rates of seeds were well maintained in PICS bags. The damaged seeds percentage and the 100 seeds weights did not show any difference from what was noted at the beginning of the experiment. Further aflatoxin tests were performed on 245 samples. In both PICS and woven bags, an aflatoxin level of above 20 ppm was recorded in 53% of the samples. The level of contamination in samples from PICS bags was much less than those from woven bags. So maize grains can be effectively stored in PICS bags in areas with a high incidence of *P. truncatus,* but to reduce bag damage due to *P. truncatus* the maize grains should be packed in PICS bags immediately after harvest and drying (Baoua et al., 2014).

The concentration of aflatoxins (AFs) B_1, B_2, G_1, and G_2 and the impact of weather conditions on the production of AFs was ascertained in maize samples collected from different regions of Serbia. The samples were subjected to direct competitive ELISA to find out the AFs content. During 2009–2011, AFs were not found in maize samples. But in 2012, the variations in weather condition resulted in the production of AFs in 137, namely, 68.5% of maize samples. These results indicated a high effect of hot and dry weather with extended drought during spring and summer in 2012 on AFs contamination in maize. This information could be utilized in the establishment of AFs monitoring programs (Kos et al., 2013).

Computers and electronics have made a specific impact on the postharvest industry which includes environmental control and storage, quality monitoring, quality management, grading systems, inventory control, and management of products. This study illustrated the present status of postharvest technology and presented some future estimates. It is expected that demand for enhanced quality, longer storage life, and assured product safety will continue to grow. The industry that meets these demands and electronic technology will have a significant role in this. The improved sensors to evaluate quality are still desirable, and handling and storage systems are expected to become highly sophisticated. Though technology has contributed much to increase the world's food supply, further it created problems for the society, which need to be solved in the next millennium (Studman, 2001).

Maize grains harvested under humid and warm conditions are susceptible to molds and undergo rapid deterioration. So, grains should be dried to safe moisture levels before storage to protect them from microbial damage. But in developing countries, the drying of produce to these moisture levels may not be economical for farmers. In such situations, preservation of grains under hermetic storage at intermediate moisture levels could be feasible and economical. In this study, the influence of different moisture contents (MC) on the quality of maize grains was examined in self-regulated modified atmospheres during hermetic storage. Initially, maize was conditioned at 14%, 16%, 18%, 20%, and 22% MC. for 28 days in tightly wrapped plastic bags and later stored in sealed containers at 30°C for 75 days. The oxygen in bags was replaced with CO_2. The reduction in O_2 depletion time, namely, from 600 h at 14% MC to 12 h at 22% was observed with the increment in MC. The maximum DM losses, the lowest germination rates and the highest yeast and bacteria counts were recorded in maize at 20% and 22% MC. In the hermetically sealed maize, the major fermentation product was ethanol (0–5 g kg^{-1} DM), together with acetic acid (0–1 g kg^{-1} DM) at low concentrations. *In vitro* experiment findings indicated that maize could be stored satisfactorily at the tested moisture levels under sealed conditions where self-regulated atmospheres protect the produce from microbial damage. However, the economic feasibility of storing high-moisture maize will need to be tested in further large-scale trials (Weinberg et al., 2008).

9.10 MARKETING THE HARVESTS

To find out a suitable strategy that can raise the adoption level of maize growers a study was conducted on the adoption and marketing behavior of maize farmers. In Tamil Nadu, three blocks of Coimbatore district, namely, Udumalpet, Pollachi, and Palladam were selected for the study. The analysis of the extent of adoption revealed that the majority of the farmers favor Adipattam season for maize sowing. Most of the farmers followed the maize cultivation practices such as basal urea application

(61.11%), top dressing (60.00%), application of micronutrient (46.70%), seed treatment (34.50%), weed crop protection (78.90%), and water management (74.40%). The produce was harvested at the dry and hard seed stage. The marketing behavior analysis has shown that the farmers prefer to sell their produce in nearby towns by transporting through tractors. Further, it was found that half of the farmers sell their entire produce through wholesalers. Before marketing the produce most of the framers performed grading and weighing but did not follow storage pest protection measures. For the majority of the maize farmers, the neighbor farmers were the key source of information. The key characteristics affecting the marketing behavior of the maize farmers were educational status, socioeconomic status, extension agency contact, storage facilities, market awareness, and market potential indicators (Jaisridhar et al., 2012).

The marketing efficiency of different maize marketing channels was determined by collecting primary data directly from 55 farmers and intermediaries from the Gaibandha district of Bangladesh. (1) Farmers–farias–wholesalers–aratdars–feed mills, (2) farmers–wholesalers–aratdars–feed mills; (3) farmers–aratdars–feed mills; (4) farmers–wholesalers–feed mills; and (5) farmers–farias–aratdars–feed mills were identified as major maize marketing channels in Bangladesh. Among these channels, channel 3, namely, farmers–aratdars–feed mill was found to be most efficient. The best alternative for this channel was channel 4 (farmers–wholesalers–feed mills). Therefore, the study emphasized that a direct buying and selling system should be developed by reducing the number of intermediaries to improve the marketing efficiency of maize (Golam Kausar and Alam, 2016).

9.11   RESEARCH ADVANCES IN CULTIVATION OF MAIZE

At present, conservation agriculture is becoming common in maize-based farming systems. Conventional agriculture involves practices such as reduced tillage, permanent soil cover, and crop rotations to improve soil fertility and to provide food from a declining land resource. Therefore, the factors influencing crop yield under conservation agriculture and rain-fed conditions need to be identified. In this chapter, the influence of long-term tillage and retention of residue on maize grain yield was assessed under different soil textures, nitrogen input, and climatic conditions. Stability analysis was used to measure the yield variability. In low rainfall areas, increased maize yield was noted with the conservation agriculture practices such as rotation and high input use, but the stability system did not show any variation under low rainfall areas. Annual rainfall showed a strong association with maize grain yield. The metaanalysis results revealed that (i) mulching in high rainfall areas may reduce yields because of water-logging; (ii) for the temporal development of conservation agriculture effects soil texture is important and well-drained soils improves maize yield; (iii) for improved yield conservation agriculture practices need high inputs particularly N; (iv) reduced tillage without mulching lowers yield in semiarid areas; (v) when sufficient fertilizer is available, rainfall is the key determinant of yield in southern Africa. Thus, the study findings clearly indicated that to improve the impact of conservation agriculture on yield it should be targeted and adapted to specific biophysical conditions (Rusinamhodzi et al., 2011).

Earlier studies provided the indication of climate change in sub-Saharan Africa; therefore, more climate-resilient maize systems need to be developed. In maize systems, the adaptation strategies to climate change include improved germplasm with tolerance to drought and heat stress and improved management practices. But to find out agricultural responses to future climate change and to set priorities for adaptation approaches, the ability to precisely estimate future climate scenarios is required. In this regard, 19 global climate models were used to predict the climate change scenarios for maize growing regions of Africa. It was predicted that throughout maize mega-environments within sub-Saharan Africa air temperatures may increase by an average of 2.1°C by 2050. During the maize growing season, rainfall fluctuations differed with location. Owing to the gap between the development of improved cultivars till the seed is in the hands of farmers and the implementation of new management practices, there is a crucial need to develop germplasm resistant to climate change to offset the projected yield declines (Cairns et al., 2013).

Crop simulation models can be applied is to assess crop yield during the current growing season. Numerous studies have tried to combine crop simulation models with remotely sensed data through data-assimilation methods. This method allows reinitialization of model parameters with remotely sensed observations to improve the performance of the model. Here, corn yield in the state of Indiana, USA was estimated by integrating the cropping system model–CERES–maize with the moderate resolution imaging spectroradiometer (MODIS) and LAI products. For most of the regions, assessed corn yield in 2000 equated relatively well with the US Department of Agriculture statistics. Planting, emergence and maturation dates, and N fertilizer application rates were also assessed at a regional level. When compared with only the green-up LAIs or the highest LAI values, the seasonal LAI in the optimization method provided the best results. Additional studies should be carried out to investigate model uncertainties by means of other MODIS products, such as the improved vegetation index (Fang et al., 2008).

9.12 CONCLUSIONS

Various phases of maize cultivation were presented in this chapter. These comprised land preparation, methods of sowing, water management, fertilizer management, harvesting storage, and marketing. In addition, research advances in maize cultivation were discussed.

KEYWORDS

- maize
- cultivation
- irrigation
- physiological maturity
- harvesting
- marketing

REFERENCES

Afuakwa, J. J.; Crookston, R. K. Using the Kernel Milk Line to Visually Monitor Grain Maturity in Maize. *Crop Sci.* **1984,** *24*, 687–691.

Baoua, I. B.; Amadou, L.; Ousmane, B.; Baributsa, D.; Murdock, L. L. PICS Bags for Post-harvest Storage of Maize Grain in West Africa. *J. Stored Prod. Res.* **2014,** *58*, 20–28.

Bennie, A. T. P.; Botha, F. J. P. Effect of Deep Tillage and Controlled Traffic on Root Growth, Water-Use Efficiency and Yield of Irrigated Maize and Wheat. *Soil Tillage Res.* **1986,** *7*, 85–95.

Bonelli, L. E.; Monzon, J. P.; Cerrudo, A.; Rizzalli, R. H.; Andrade, F. H. Maize Grain Yield Components and Source–Sink Relationship as Affected by the Delay in Sowing Date. *Field Crops Res.* **2016**, *198*, 215–225.

Brooking, I. R. Maize Ear Moisture during Grain-Filling, and Its Relation to Physiological Maturity and Grain-Drying. *Field Crops Res.* **1990**, *23*, 55–68.

Bruns, H. A. Controlling Aflatoxin and Fumonisin in Maize by Crop Management. *J. Toxicol. Toxin Rev.* **2003**, *22*, 153–173.

Cairns, J. E.; Hellin, J.; Sonder, K.; Araus, J. L.; MacRobert, J. F.; Thierfelder, C.; Prasanna, B. M. Adapting Maize Production to Climate Change in Sub-Saharan Africa. *Food Sec.* **2013**, *5*, 345–360.

Chaudhary, M. R.; Prihar, S. S. Root Development and Growth Response of Corn Following Mulching, Cultivation, or Interrow Compaction. *Agron. J.* **1974**, *66*, 350–355.

de Oliveira, I. J.; Atroch, A. L.; Dias, M. C.; Guimarães, L. J.; de Oliveira Guimarães, P. E. Selection of Corn Cultivars for Yield, Stability, and Adaptability in the State of Amazonas, Brazil. *Pesq. Agropec. Bras.* **2017**, *52*, 455–463.

Fang, H.; Liang, S.; Hoogenboom, G.; Teasdale, J.; Cavigelli, M. Corn-Yield Estimation Through Assimilation of Remotely Sensed Data into the CSM-CERES-Maize Model. *Int. J. Remote Sens.* **2008**, *29*, 3011–3032.

Gasura, E.; Setimela, P.; Edema, R.; Gibson, P. T.; Okori, P.; Tarekegne, A. Exploiting Grain-Filling Rate and Effective Grain-Filling Duration to Improve Grain Yield of Early-Maturing Maize. *Crop Sci.* **2013**, *53*, 2295–2303.

Grassini, P.; Thorburn, J.; Burr, C.; Cassman, K. G. High-Yield Irrigated Maize in the Western U.S. Corn Belt: I. On-Farm Yield, Yield Potential, and Impact of Agronomic Practices. *Field Crops Res.* **2011**, *120*, 142–150.

Golam Kausar, A. K. M.; Alam, M. J. Marketing Efficiency of Maize in Bangladesh. *Asian J. Agric. Ext. Econ. Sociol.* **2016**, *11*, 1–12.

Gu, R.; Li, L.; Liang, X.; Wang, Y.; Fan, T.; Wang, Y.; Wang, J. The Ideal Harvest Time for Seeds of Hybrid Maize (*Zea mays* L.) XY335 and ZD958 Produced in Multiple Environments. *Sci. Rep.* **2017**, *7*, 17537.

Jaisridhar, P.; Ravichandran, V.; Jadoun, Y. S.; Senthil Kumar, R. Study on Adoption and Marketing Behaviour of Maize Growers in Viour of Maize Growers in Coimbatore District of Tamil Nadu. *Indian J. Agric. Res.* **2012**, *46*, 173–177.

Janardhana, G. R.; Raveesha, K. A.; Shetty, H. S. Mycotoxin Contamination of Maize Grains Grown in Karnataka (India). *Food Chem. Toxicol.* **1999**, *37*, 863–868.

Kos, J.; Mastilović, J.; Hajnal, E. J.; Šarić, B. Natural Occurrence of Aflatoxins in Maize Harvested in Serbia during 2009–2012. *Food Contr.* **2013**, *34*, 31–34.

Kumar, P.; Kumar, M.; Kumar, R.; Upadhaya, B.; Hussain, M. Z; Raushan, R. K. Effect of Nutrient Management on Nutrient Uptake and Economics of Maize (*Zea mays* L.) under Different Tillage Practices. *Int. J. Curr. Microbiol. Appl. Sci.* **2019**, *8*, 783–789.

Lone, A. A.; Allai, B. A.; Nehvi, F. A. Growth, Yield and Economics of Baby Corn (*Zea mays* L.) as Influenced by Integrated Nutrient Management (INM) Practices. *Afr. J. Agric. Res.* **2013**, *8*, 4537–4540.

Roja, M.; Kumar, K. S.; Ramulu, V.; Satish, C. Water productivity of Rabi Maize Influenced by Different Drip Irrigation Treatments. *Int. J. Agric. Sci.* **2017**, *9*, 4515–4517.

Rusinamhodzi, L.; Corbeels, M.; van Wijk, M. T.; Rufino, M. C.; Nyamangara, J.; Giller, K. E. A Meta-Analysis of Long-Term Effects of Conservation Agriculture on Maize Grain Yield under Rain-Fed Conditions. *Agron. Sustain. Dev.* **2011,** *31,* 657–673.

Saha, S.; Gopinath, K. A.; Mina, B. L.; Gupta, H. S. Influence of Continuous Application of Inorganic Nutrients to a Maize–Wheat Rotation on Soil Enzyme Activity and Grain Quality in a Rainfed Indian Soil. *Eur. J. Soil Biol.* **2008,** *44,* 521–531.

Sala, R. G.; Andrade, F. H.; Westgate, M. E. Maize Kernel Moisture at Physiological Maturity as Affected by the Source–Sink Relationship during Grain Filling. *Crop Sci.* **2007,** *47,* 711–716.

Singh, A. Study of Different Methods of Maize Harvesting and Threshing in Hoshiarpur District of Punjab. *Int. J. Agric. Eng.* **2014,** *7,* 267–270.

Studman, C. J. Computers and Electronics In Postharvest Technology—A Review. *Comput. Electron. Agric.* **2001,** *30,* 109–124.

Tefera, T. Post-harvest Losses in African Maize in the Face of Increasing Food Shortage. *Food Sec.* **2012,** *4,* 267–277.

Tiwari, D. K.; Hooda, V. S.; Thakral, S. K.; Yadav, A.; Sharma, M. K. Effect of Planting Methods, Maize Hybrids and Nitrogen Levels on Nutrient Uptake of High-Quality Protein Maize (*Zea mays* L.). *Int. J. Curr. Microbiol. Appl. Sci.* **2018,** *7,* 246–253.

Tolera, A.; Sundstøl, F.; Said, A. N. The Effect of Stage of Maturity on Yield and Quality of Maize Grain and Stover. *Anim. Feed Sci. Technol.* **1998,** *75,* 157–168.

Tsimba, R.; Edmeades, G. O.; Millner, J. P.; Kemp, P. D. The Effect of Planting Date on Maize Grain Yields and Yield Components. *Field Crops Res.* **2013,** *150,* 135–144.

Valentinuz, O. R.; Tollenaar, M. Vertical Profile of Leaf Senescence during the Grain-Filling Period in Older and Newer Maize Hybrids. *Crop Sci.* **2004,** *44,* 827–834.

Velu, V.; Nagender, A.; Prabhakara Rao, P. G.; Rao, D. G. Dry Milling Characteristics of Microwave Dried Maize Grains (*Zea mays* L.). *J. Food Eng.* **2006,** *74,* 30–36.

Weinberg, Z. G.; Yan, Y.; Chen, Y.; Finkelman, S.; Ashbell, G.; Navarro, S. The Effect of Moisture Level on High-Moisture Maize (*Zea mays* L.) Under Hermetic Storage Conditions—*In Vitro* Studies. *J. Stored Prod. Res.* **2008,** *44,* 136–144.

Zhang, X.; Li, H.; He, J.; Wang, Q.; Golabi, M. H. Influence of Conservation Tillage Practices on Soil Properties and Crop Yields for Maize and Wheat Cultivation in Beijing, China. *Austr. J. Soil Res.* **2009,** *47,* 362–371.

CHAPTER 10

Maize Grain Quality Analysis, Food Quality, Chemistry, and Food Processing

ABSTRACT

Studies on the improvement of maize quality are carried out through various methods and many advancements have been attained that need to be gathered at one place. Therefore, in this chapter, recent methods and their effectiveness in the determination of grain characteristics and food quality are being presented. A brief overview of the recent advancements made in maize grain quality research is given.

10.1 MAIZE GRAIN QUALITY ANALYSIS

A traditional Ghanaian fermented maize food kenkey was prepared by supplementing it with cowpea-mediated protein. In the laboratory, kenkey was made by mixing maize and red or white cowpea in a 4:1 ratio and compared with other maize products for parameters like the dough fermentation profile, color, and fracture profiles. The products were fermented at 30°C and after 4 days significant differences were not observed in fermentation profiles of the products. The final dough pH values for maize and maize/cowpea mixtures were 4.07 and 4.08, respectively. The kenkey supplemented with 20% white and 20% red cowpea showed 12·99% (w/w) and 13·89% (w/w; dry weight basis) increment in crude protein content. But, the whiteness viz. Hunter L value of kenkey was reduced with the addition of white cowpeas (12%) and red cowpeas (27%). When compared with traditional maize kenkey, the kenkey supplemented with cowpea was more homogeneous and less susceptible to fracture. The force needed to fracture kenkey supplemented with cowpea was higher than for traditional kenkey and was found to increase with rising concentrations of cowpea. A group of native Ghanaians acquainted with the traditional maize kenkey was made to test the kenkey supplemented with cowpea, and they confirmed that the quality parameters of the new product were comparable or even better than traditional kenkey (Nche et al., 1994).

The effects of solid (5–13%) and sugar (0–54%) contents on the viscosity of a semi-liquid maize food, *akamu* were investigated. The viscosity of *akamu* was determined at 30, 40, 50, 60, and 70°C using a rotational viscometer at speeds between 10 and 100 rev/min. *Akamu* showed a pseudoplastic behavior with a power law index value of 0.32. Temperature, solid content, and sugar content did not show any effect on pseudoplastic behavior. However, a reduction in consistency indices was observed with temperature and sugar content, but it increased with solid content. These dependencies were described by using exponential equations. Solid content had a higher effect on viscosity and sweetened *akamu* was found to be more sensitive to variations in temperature (Sopade and Filibus, 1995).

An experiment was performed to study the effects of N application on maize (*Zea mays* L.) grain quality parameters. Five maize cultivars were evaluated by applying four levels viz. 0, 30, 60, and 120 kg ha^{-1} of nitrogen. In all the five cultivars, an increment in grain yield, kernel weight, and grain protein was observed with an increasing level of N. Highest grain yield (5.3 Mg ha^{-1}) and kernel weight (26.62 mg) were recorded in the hybrid 8644-27. Average grain protein yield per unit area did not show a significant difference among cultivars. In comparison, the percentage of floaters was higher in the cultivars 8644-27 and TZPB-SR at 30–60 kg N ha^{-1}. This reflects a greater proportion of floury endosperm in the cultivars, showing that they would be best for traditional dry milling, where the flour is produced by grinding whole grain. Further, the cultivars SPL and TZB-SR had a low percentage of floaters and high test weights at 30–60 kg N ha^{-1} indicating that these cultivars would give high yields of grits when processed and have high value for industrial dry milling. Thus, the research findings have indicated that the selection of cultivar and N level may affect grain quality and they should be taken into account while growing maize for dry milling purposes (Oikeh et al., 1998).

The effects of different shading stages after anthesis on maize grain weight and quality were investigated at the cytology level. The stress was provided on plants at 1-14 d (SI), 15-28 d (S2), and 29-42 d (S3) after pollination by using a horizontal shading net and by reducing the light intensity to 55%. Grain weight, quality, endosperm cell proliferation, cob sugar content, and grain pedicel vascular bundle cross-section area were evaluated. After pollination, the ultrastructural variations in endosperm cells and endosperm transfer cells were noticed. The results showed reductions in the grain weight, starch content, endosperm cell number, and volume of shaded plants. On the other hand, the increment was observed in the proportion of embryo and endosperm, protein content, and fat content of shade plants grain. It was found that the shade significantly delays the starch granules development and remarkably reduces the endosperm filling status. The plants under shading stress at 1–14 d after pollination exhibited the least volume of grain endosperm. But, in plants under shading at 15–28 d after pollination, the volume of starch granules and the substantiation of endosperm were the worst. The shaded plants showed increase soluble sugar of maize cob without any noticeable change in vascular structure of small cluster stalk. The low light at different stages significantly affected the protein body number in maize endosperm. Moreover, the reductions in degree of connection, nutrient transport capacity, and mitochondrion number of the transfer cell were observed after shading. The variation in grain quality under shade was found to be due to the increase in embryo and endosperm proportion. The variations in the morphology and functions of transfer cell endosperm and the lack of energy limited the nutrient transport significantly with shading at different stages, indicating that a blocked flux may be the reason for the reduction of maize grain weight at later growth stage under low light condition (Shi-fang et al., 2011).

10.1.1 GRAIN QUALITY CHARACTERISTICS

A dramatic increase has been experienced in maize grain yield over the past 50 years, and simultaneously its end uses have increased demanding special quality traits. Many special types of maize have been developed by plant breeders, all of which are affected by the agronomic practices applied to produce the crop. A decreasing trend in protein concentration was observed with increasing grain yield, but the increase in yield resulting from nitrogen application found to increase the protein concentration. Irrigation improved the protein biological value, while high nitrogen applications change the amino acid balance thus decreasing the nutritional value. Kernel breakage susceptibility and kernel density were reduced by irrigation and increased with high nitrogen applications. Large effects of hybrid choice and small effects of production practice were recorded on extractable starch and oil concentration. Soil or foliar fertilizer application exhibited a greater effect on essential mineral nutrient levels. Irrigation and water stress management practices during grain filling reduced the production of aflatoxin. Thus, the study findings suggested that though genetics play a significant role in defining the maize grain quality, agronomic practices also exert a significant effect on it (Mason and D'croz-Mazon, 2008).

Researchers are concerned that breeding for improved yield may decline the maize grain quality. In this paper, the chemical and physical quality parameters of maize were quantified for hybrids representing three decades of grain yield improvement in Ontario, Canada. Six hybrids were planted with two plant densities at two locations in southern Ontario. N, P, K, Ca, Mg, Zn, Cu, Mn, Se, lysine, tryptophan, and lipid contents of grain were assessed as chemical quality parameters. The physical quality parameters measured were test weight, kernel density, stress cracks, and breakage susceptibility. The old hybrids had greater Mg, Cu, Mn, and Se concentrations in kernels, whereas N concentration was higher in kernels of recent hybrids. Hybrids grown at high plant density had lower levels of N, P, and Mn than those at low plant density. Among hybrids, lysine and tryptophan concentrations did not vary significantly. Variations in nutrient concentration were often greater among hybrids within an era than among means for eras of hybrid release. Physical quality parameters such as test weight, kernel density, and kernel weight were found to increase with more recent hybrids. Further, these traits were higher in hybrids grown at low plant density. Though kernels breakage susceptibility was not influenced by the era of hybrid release, it was less at low plant density (Vyn and Tollenaar, 1998).

Physicochemical, morphological, thermal, and rheological properties of popcorn, dent corn, and baby corn (dent type) grains were evaluated. The grains of these corns were fractioned into different fractions and their starches were separated. The fractions of dent corn and popcorn showed significant differences in various properties. The shape of starch granules varied from oval to polyhedral. Popcorn and dent corn starch exhibited the presence of polyhedral shape granules, while baby corn starch had oval shape granules. In starches from different corn types, the amylose content was found to vary between 15.3% and 25.1%. The lowest swelling power, solubility, amylose content, and mean granule diameter were recorded in baby corn. Then, using differential scanning calorimetry, the transition temperatures (T_o, T_p, and T_c) and gelatinization enthalpy (ΔH_{gel}) of starches were determined. T_o, T_p, T_c, and ΔH_{gel} ranged from 66.3 to 69.3, 71.5 to 73.1, 76.5 to 78 °C, and 8.9 to 10.9 J/g, respectively. The highest T_o, T_p, ΔH_{gel}, and PHI were noted in the fractions of dent corn and popcorn starch. By means of a dynamic rheometer, the rheological properties of the starches from popcorn and dent corn were measured. The rheological properties of these corns have shown a significant variation in the peak G', G'', and peak tanδ values. The highest peak G', G'' values, and breakdown in G were found in a large grain fraction of popcorn. In all the corn types, the increment in turbidity of the gelatinized aqueous starch pastes was observed with the increase in storage period. During storage highest retrogradation, values were recorded in baby corn starch, while the large grain fraction of popcorn showed the highest values (Sandhu et al., 2004).

10.2 FOOD CHEMISTRY

Maize populations from diverse origins were evaluated for their food grain quality-related traits. The calibration equations for the evaluation of crude protein, crude lipid, starch, and floatation area were developed by scanning maize samples with near-infrared (NIR) spectroscopy. The populations VA25, VA158, VA282, VA284, VA285, VA567, VA572, VA814, VA950, VA1057, and VA1179 had protein and lipid contents varying between 12.52 and 15.16% and 5.26 and 7.17%, and found promising from a nutritional viewpoint. In the EUMLCC, the range of variation noted for antioxidants was quite large. Lutein ranged from 1.03 to 21.00 mg kg^{-1} dm; zeaxanthin ranged from 0.01 to 35.00 mg kg^{-1} dm; and total carotenoids varied between 1.09 and 61.10 mg kg^{-1}dm. Recently, a single cross-hybrid was developed from the ITA0370005 population; this hybrid had a high carotenoids content and is now being used by the Italian food industry (Berardo et al., 2009).

The starches from diverse corn varieties such as African Tall, Ageti, Early Composite, Girja, Navjot, Parbhat, Partap, PbSathi, and Vijay were compared for their physicochemical, thermal, pasting, and gel textural properties. Among different corn varieties, the range for peak viscosity was 804–1252 cP. The starch gels hardness varied from 21.5 to 32.3 g. The ranges of gelatinization (ΔH_{gel}) enthalpy and retrogradation (%R) percentage for different corn starches were 11.2–12.7 J/g and 37.6–56.5%, respectively.

Amylose content ranged from 16.9% to 21.3%, while swelling power ranged from 13.7 to 20.7 g/g. The higher swelling power, peak, trough, breakdown, final and setback viscosity, and lower ΔH_{gel} and gelatinization range were recorded in African Tall and Early Composite. Gelatinization onset temperature (T_o) exhibited a negative correlation with peak, breakdown, final and setback viscosity, and a positive correlation with pasting temperature. ΔH_{gel} had a positive correlation with T_o, peak gelatinization temperature and (T_p), and gelatinization conclusion temperature T_c, while it exhibited a negative correlation with peak, breakdown, final, and setback viscosity (Sandhu and Singh, 2007).

Eighteen Mexican maize cultivars were analyzed for free-radical scavenging, reducing activity and for phytochemical contents like total phenolic, anthocyanin, and ferulic acid. In maize whole grain flour, the total phenolic content was between 215.8 and 3400.1 mg of gallic acid/100 g and total anthocyanin was 1.54 to 850.9 mg cyanidin-glucoside equivalents/100 g. In grain, most of the phenolic contents were present in bound form and anthocyanins were the major free phenolic compounds. Maximum antiradical and reducing activities were recorded in bound phenolic extracts of corn. The cultivars Veracruz 42 and AREQ516540TL were purple in color with enriched anthocyanins, and showed the greatest antiradical and reducing activities. The red-colored cultivar, Pinto exhibited the highest antiradical activities. Therefore, variations in free-radical scavenging and reducing activities were found to be reliant on the anthocyanins and other phenolics unique profile in each cultivar (Lopez-Martinez et al., 2009).

In this regard, the applicability of genetically modified (GM) maize and soy quantification to processed foods was examined with the help of heat treatment processing models. The detection methods were based on real-time quantitative polymerase chain reaction (PCR) analysis. Ground seeds of insect resistant GM maize (MON810) and glyphosate tolerant Roundup Ready (RR) soy were dissolved in water and autoclaved at different time intervals. In the extracted DNA solution, reduction in the recombinant copy numbers and taxon specific deoxyribonucleic acid (DNA) sequences was observed with time. This reduction was affected by the PCR-amplified size. The ratio of the recombinant DNA sequence to the taxon-specific DNA sequence, i.e. conversion factor (C_f) was used as a constant number to calculate GM% at each event. When the sizes of two DNA sequences PCR products were nearly equal, GM% tended to be stable. This indicated that PCR product size plays a main role in quantifying GM organisms in processed foods. It was found that the origin of GM from a paternal or maternal source influences the C_f of the endosperm ($3n$). From the F_1 generation seeds of five GM maize, the embryos and endosperms were separated and their C_f values were calculated. Both paternal and maternal GM events were detected. In these events, C_f of endosperm was lower than that of the embryo, and the C_f of the embryo was lower than that of the endosperm. These findings demonstrated the problems encountered while determining GM% in F_2 generation maize grains and in processed foods from maize and soy (Yoshimura et al., 2005).

Although maize supplies macro- and micronutrients essential for humans, it is deficient in essential amino acids lysine and tryptophan. People who consume >50% of their daily energy from maize may suffer from protein malnutrition. Severe malnutrition of protein and energy raises susceptibility to life-threatening diseases like tuberculosis and gastroenteritis. A nutritionally superior maize cultivar, quality protein maize (QPM) has double the amount of lysine and tryptophan, and protein bioavailability that equivalents milk casein. Studies on animals and humans suggested that substitution of common maize with QPM can improve health. But, the practical contribution of QPMs to maize-subsisting populations is still unsolved.

In this study, the WHO and Institute of Medicine recommendations of total protein and essential amino acid were used to assess QPM target consumption levels for young children and adults, and these were compared with average daily maize consumptions by African countries. The results unveiled that to maintain the sufficiency of lysine, the limiting amino acid in maize ~100 g QPM is essential for children, and about 500 g is essential for adults. This represented a 40% decrease in maize consumption compared with common maize to meet protein needs. In Africa, the significance of maize highlights the potential for QPM to aid in closing the protein insufficiency gap (Nuss and Tanumihardjo, 2011).

To foresee the functional properties of products resulting from the whole maize flour (nDF) and dehulled-degermed maize flour (DF), their physicochemical properties were examined. The yield of flour from whole milled grains was 92%, while it was only 65% from dehulled-degermed grains. Reduced levels of crude protein (8.2–7.5 %), fat (6.0–1.4 %), crude fiber (1.8–0.5 %), and ash (1.5–0.2 %) were recorded in flour produced by degerming and dehulling. DF had higher carbohydrates and water absorption capacity than nDF. Therefore, the swelling capacity of DF was about 1.5 times that of nDF. Brabender hot-paste viscometry revealed that total solids slurry (DF) of 10% was sufficient to give about 500 BU consistency. The average peak hot-paste viscosity for DF was 505 BU, and for nDF it was 370 BU. Thus, DF products would be more sticky and pasty than those of nDF. Starch, fat, and fiber contents were found to be the major factors leading to variances in functional properties shown by the DF and nDF flours (Houssou and Ayernor, 2002).

10.3   FOOD PROCESSING

An investigation was carried out to study the effects of food processing and maize variety on amounts of starch escaping digestion in the small intestine. In this study, five subjects with ileostomies were fed with two meals differing in resistant starch (RS) concentration, but having the same macronutrient composition and nonstarch polysaccharides. The high-RS meal comprised bread prepared from high-amylose maize, uncooked green banana flour, and coarsely ground uncooked wheat. The low-RS meal included bread prepared from low-amylose maize, cooked green banana flour, and cooked wheat. The total amount of starch escaping digestion was analyzed in effluent produced over 14 h. In the low-RS meal, from the 51.8 ± 6.2 g of starch fed 2.4 ± 0.6 g was recovered in the effluent. In the high-RS meal, starch consumed was 52.7 ± 8.8 g of which 19.9 ± 5.2 g recovered in the effluent. The effluent scanning electron micrographs from one subject fed with high-amylose bread unveiled that several intact starch granules escaped digestion in the small intestine (Muir et al., 1995).

The bioavailability of micronutrients is one of the aspects of dietary quality related with the adequacy of nutrition. In plant-based diets, the bioavailability of micronutrients can be improved by using traditional household food-processing and preparation methods. These methods comprise thermal processing, mechanical processing, soaking, fermentation, and germination/malting. The objective of these strategies is to improve the physicochemical availability of micronutrients, reduce the content of antinutrients, like phytate, or increase the content of compounds that improve bioavailability. Further, to ensure a positive and significant effect on the adequacy of micronutrient, a combination of different strategies is required. In Malawi, a long-term participatory intervention used these strategies along with the promotion of other micronutrient-rich foods consumption, such as animal-source foods. When compared with control children, this intervention resulted in improvements in both hemoglobin and lean body mass and a lower occurrence of common infections. However, these strategies need to be broadly evaluated for their suitability and effect on nutritional status (Hotz and Gibson, 2007).

The increasing use and consumption levels of maize in developing countries demand new methods of processing to improve the functionality and nutrient quality of maize-based foods. In this paper, the effect of nixtamalization on the chemical and functional characteristics of maize was determined. Using standard methods chemical composition viz., moisture, protein and ash contents, pH, titratable acidity, water absorption, color, cooked paste viscosity, and texture were determined. Significant influence of cooking time and lime concentration was observed on the moisture, pH, and color of the maize samples. It was found that the water absorption ability depends on the lime concentration and an increment in all the indices was observed with the increasing concentration of lime. However, the lime concentration did not show any significant effects on cooked paste viscosity, ash, and protein contents. Therefore, to make snack products, maize nixtamal with suitable moisture and texture can be prepared from maize with acceptable quality traits (Sefa-Dedeh et al., 2004).

The common mycotoxins associated with cereal grains are aflatoxin, ochratoxin A, fumonisin, deoxynivalenol, and zearalenone. These mycotoxins can not be eliminated completely during food processing. But, various food processing operations like sorting, trimming, cleaning, milling, brewing, cooking, baking, frying, roasting, canning, flaking, alkaline cooking, nixtamalization, and extrusion may reduce their concentration in processed foods. Though it needs very high temperatures to reduce mycotoxin concentrations, it was found that roasting and extrusion processing has the potential to reduce mycotoxin concentrations. To get a good level of reduction in zearalenone and fumonisin, moderate reduction of aflatoxin, and variable to low reduction of deoxynivalenol, the extrusion processing has to be performed at temperatures above 150 °C. At extrusion temperatures of 160 °C or above and in the presence of glucose, maximum reductions of fumonisin were recorded. The extrusion processing with 10% added glucose reduced the levels of Fumonisin B_1 by 75–85%. In a feeding trial, the rats fed with extruded fumonisin contaminated corn grits showed some reduction in toxicity of grits extruded with glucose (Bullerman and Bianchini, 2007).

This study described the most promising method of high resistant starch product preparation. Here, the high starch product was made by hydrolyzing maize starch with pullulanase. The maize starch was hydrolyzed at 46°C for 32 h by adding 12 ASPU/g pullulanase. Then, the hydrolysate obtained by pressure-cooking was autoclaved at 121°C for 1 h, cooled at room temperature, and stored at 4°C overnight. The autoclaving and cooling were repeated for two cycles. After, drying in an oven at 105°C, it was ground into fine particles (<150 μm). The resistant starch content in the final product was 44.7% (w/w) (Zhang and Jin, 2011).

The protein content of corn is 45–50% and the major storage protein is zein. Based on the zein solubility in aqueous and alcohol solutions, it was first identified in 1897. Because of the negative nitrogen balance and poor solubility of zein isolate in water, it is not used directly for human consumption. The present production of zein from corn gluten meal is restricted to ≈500 tonnes per year. The capacity of zein and its resins to form tough, glossy, hydrophobic grease-proof coatings resistant to microbial attack made it a product of commercial interest. Depending on its purity, zein sells for ≈US$10–40 per kilogram. The use of zein in fiber, adhesive, coating, ceramic, ink, cosmetics, textile, chewing gum, and biodegradable plastics are some of its potential applications. These novel uses of zein seem promising, but low-cost manufacturing methods will need to be developed (Shukla and Cheryan, 2001).

The loss of vitamin C during processing is the main reason for the lower nutritional value of processed fruits and vegetables than fresh produce. It has been found that in apples vitamin C contributes <0.4% of total antioxidant activity, which indicates that the natural combination of phytochemicals is responsible for most of the antioxidant activity. This suggested that in spite of the vitamin C loss processed fruits and vegetables may contain antioxidant activity. This study reported that the thermal processing of sweet corn at 115 °C for 25 min increased the total antioxidant activity by 44% and phytochemical content like ferulic acid by 550% and total phenolics by 54%, though vitamin C loss of 25% was noted. Processed sweet corn had improved antioxidant activity equal to 210 mg of vitamin C/100 g of corn compared to the remaining 3.2 mg of vitamin C in the sample that added only 1.5% of its total antioxidant activity. Their results were not in agreement with the view that processed fruits and vegetables contain lower nutritional value than fresh produce. However, this report may have a substantial influence on consumers' food selection by increasing their fruits and vegetables consumption so as to lessen the risk of chronic diseases (Dewanto et al., 2002).

The antimicrobial effects of the various processes involved in the fermented maize dough porridge preparation were evaluated. In the laboratory, fermented maize dough porridge was inoculated with *Shigella flexneri* and enterotoxigenic *Escherichia coli* (ETEC), and unhygienic conditions of a typical rural community were created. The reduction in the pH and no antimicrobial effect against *Shigella* and ETEC were observed during the soaking process. Unfermented maize dough did not show any inhibition effect on test strains. The fermented maize dough inhibited the test strains after 8 h of inoculation. A significant reduction in the anti-microbial effect was observed in porridge cooked from the fermented maize dough. These findings suggested that the fermentation of maize dough could be used to reduce the contamination of weaning foods by *Sh. flexneri* and ETEC (Mensah et al., 1991).

Two major mycotoxins that co-occur in maize are aflatoxins and fumonisins. This study investigates the fate of these mycotoxins during the traditional processing. Levels of mycotoxins in maize-based foods such as *mawe, makume, ogi, akassa,* and *owo were measured during* processing operations. Significant reductions in mycotoxin levels were observed. The mycotoxin reductions were higher during the preparation of makume (93% reduction of aflatoxins and 87% reduction of fumonisins) and akassa (92% reduction of aflatoxins and 50% reduction of fumonisins) than that of owo (40% reduction of aflatoxins and 48% reduction of fumonisins). The unit operations of processing like sorting, winnowing, washing, and crushing combined with dehulling were found to be very effectual in decreasing mycotoxins. The cooking and fermentation had very little effect on myco-toxins (Fandohan et al., 2005).

10.4 IMPROVING OR MAINTAINING MAIZE QUALITY

The nutritional superiority of QPM maize has been proved in nutritional trials on human and livestock consumption. In this article, the recent advancements in the development of QPM varieties with the help of conventional and molecular methods under stressed and nonstressed conditions were reviewed. Many breeding approaches have been conven-tionally applied to develop QPM varieties tolerant to stress. Three genetic loci were involved in regulating the protein synthesis factor levels associated with lysine. These loci were mapped on chromosomes 2, 4, and 7. More emphasis should be given to the development of QPM varieties tolerant to abiotic stress and post-harvest losses should also be taken into account. The minimum cooperation between maize breeders, farmers, and agricultural extension workers has been found to be the main reason for the limited adoption of QPM varieties by the farmers. So, there is a need to use a participatory plant breeding approach to improve and increase the adoption of QPM varieties.

To increase quality protein intakes especially among young children, quality protein maize (QPM) varieties are biofortified or nutritionally improved, which have high lysine and tryptophan levels. In this study, the sufficiency of children's protein consumptions in Ethiopia, where QPM is being promoted to replace conventional maize in diets, was assessed. In rural southern Ethiopia, diets of randomly sampled children aged between 12 and 36 months were measured after harvest during relative food security and 3–4 months later during relative food insecurity using 24-h weighed food records. Using the protein digestibility corrected amino acid score (PDCAAS) method, diets were analyzed for protein sufficiency, and potential improvements from QPM substitution were assessed.

At the first assessment, stunting was found to be predominant (38%). Across seasons, 95–96% of children consumed maize, which supplied 59–61% of energy and 51–55% of total protein in 24 h. Dietary consumption reduced in the food insecure season, though children were older. Among children no longer breastfeeding, QPM was likely to decrease insufficiency of utilizable protein intakes from 17% to 13% in the food secure season and from 34% to 19% in the food insecure season. But, breastfed children had only 4–6% insufficient consumption of utilizable protein, limiting the effect of QPM's potential. Owing to small farm sizes, maize stores from home production lasted a median of 3 months. Young Ethiopian children were at risk of insufficient quality protein consumptions, mainly after breastfeeding has stopped and during food insecurity. QPM could lessen this risk; however, dependence on access through home production may bring about only short-term benefits given the inadequate quantities of maize produced and stored (Gunaratna et al., 2019).

10.5 CONCLUSIONS

To improve maize quality, the grain quality traits determination is essential. In this chapter, outlines of the different techniques that can be utilized to ascertain the physical and chemical characteristics of maize quality and the processing methods to improve grain quality are presented. Further, the research advancements made to improve the maize quality are reviewed.

KEYWORDS

- **maize**
- **grain quality**
- **food quality**
- **breeding techniques**

REFERENCES

Berardo, N.; Mazzinelli, G.; Valoti, P.; Laganà, P.; Redaelli, R. Characterization of Maize Germplasm for the Chemical Composition of the Grain. *J. Agric. Food Chem.* **2009**, *57*, 2378–2384.

Bullerman, L. B.; Bianchini, A. Stability of Mycotoxins During Food Processing. *Int. J. Food Microbiol.* **2007**, *119*, 140–146.

Dewanto, V.; Wu, X.; Liu, R. H. Processed Sweet Corn has Higher Antioxidant Activity. *J. Agric. Food Chem.* **2002**, *50*, 4959–4964.

Fandohan, P.; Zoumenou, D.; Hounhouigan, D. J.; Marasas, W. F. O.; Wingfield, M. J.; Hell, K. Fate of Aflatoxins and Fumonisins During the Processing of Maize into Food Products in Benin. *Int. J. Food Microbiol.* **2005**, *98*, 249–259.

Gunaratna, N. S.; Moges, D.; De Groote, H. Biofortified Maize can Improve Quality Protein Intakes Among Young Children in Southern Ethiopia. *Nutrients* **2019**, *11*, 192.

Hotz, C.; Gibson, R. S. Traditional Food-Processing and Preparation Practices to Enhance the Bioavailability of Micronutrients in Plant-Based Diets. *J. Nutr.* **2007**, *137*, 1097–1100.

Houssou, P.; Ayernor, G. S. Appropriate Processing and Food Functional Properties of Maize Flour. *Afr. J. Sci. Technol.* **2002**, *3*, 126–131.

Lloyd, B.; Bullerman; Bianchini, A. Stability of Mycotoxins During Food Processing. *Int. J. Food Microbiol.* **2007**, *119*, 140–146.

Lopez Martinez, L. X.; Oliart-Ros, R. M.; Valerio-Alfaro, G.; Lee, C. H.; Parkin, K. L.; Garcia, H. S. Antioxidant Activity, Phenolic Compounds and Anthocyanins Content of Eighteen Strains of Mexican Maize. *LWT - Food Sci. Technol.* **2009**, *42*, 1187–1192.

Mason, S. C.; D'croz-Mazon, N. E. Agronomic Practices Influence Maize Grain Quality. *J. Crop Prod.* **2002**, *5*, 75–91.

Mensah, P.; Tomkins, A. M.; Drasar, B. S.; Harrison, T. J. Antimicrobial Effect of Fermented Ghanaian Maize Dough. *J. Appl. Bacteriol.* **1991**, *70*, 203–210.

Muir, J. G.; Birkett, A.; Brown, I.; Jones, G.; O'Dea, K. Food Processing and Maize Variety Affects Amounts of Starch Escaping Digestion in the Small Intestine. *Am. J. Clin. Nutr.* **1995**, *61*, 82–89.

Nche, P. F.; Nout, M. J. R.; Rombouts, F. M. The Effect of Cowpea Supplementation on the Quality of Kenkey, A Traditional Ghanaian Fermented Maize Food. *J. Cereal Sci.* **1994**, *19*, 191–197.

Nuss, E. T.; Tanumihardjo, S. A. Quality Protein Maize for Africa: Closing the Protein Inadequacy Gap in Vulnerable Populations. *Adv. Nutr.* **2011**, *2*, 217–224.

Oikeh, S. O.; Kling, J. G.; Okoruwa, A. E. Nitrogen Fertilizer Management Effects on Maize Grain Quality in the West African Moist Savanna. *Crop Sci.* **1998**, *38*, 1056–1161. DOI:10.2135/cropsci1998.0011183X003800040029x\

Sefa-Dedeh, S.; Cornelius, B.; Sakyi-Dawson, E.; Ohene Afoakwa, E. Effect of Nixta-malization on the Chemical and Functional Properties of Maize. *Food Chem.* **2004**, *86*, 317–324.

Shi-fang, J. I. A; Cong-feng, L. I.; Shu-ting, D.; Ji-wang, Z. Effects of Shading at Different Stages After Anthesis on Maize Grain Weight and Quality at Cytology Level. *Agric. Sci. China.* **2011**, *10*, 58–69.

Shukla, R.; Cheryan, M. Zein: The Industrial Protein from Corn. *Ind. Crops Prod.* **2001**, *13*, 171–192.

Sandhu, K. S.; Singh, N. Some Properties of Corn Starches II: Physicochemical, Gelatinization, Retrogradation, Pasting and Gel Textural Properties. *Food Chem.* **2007**, *101*, 1499–1507.

Sandhu, K. S.; Singh, N.; Kaur, M. Characteristics of the Different Corn Types and Their Grain Fractions: Physicochemical, Thermal, Morphological, and Rheological Properties of Starches. *J. Food Eng.* **2004**, *64*, 119–127.

Sopade, P. A.; Filibus, T. E. The Influence of Solid and Sugar Contents on Rheological Characteristics of Akamu, A Semi-Liquid Maize Food. *J. Food Eng.* **1995**, *24*, 197–211.

Tanzi, N. L.; Mutengwa, C. S.; Ngonkeu, E. L. M.; Gracen, V. Breeding for Quality Protein Maize (QPM) Varieties: A Review. *Agronomy* **2017**, *7*, 80–89.

Vyn, T.J.; Tollenaar, M. Changes in Chemical and Physical Quality Parameters of Maize Grain During Three Decades of Yield Improvement. *Field Crops Res.* **1998**, *59*, 135–140.

Yoshimura, T.; Kuribara, H.; Matsuoka, T.; Kodama, T.; Iida, M.; Watanabe, T.; Akiyama, H.; Maitani, T.; Furuis, S.; Hino, A. Applicability of the Quantification of Genetically Modified Organisms to Foods Processed from Maize and Soy. *J. Agric. Food. Chem.* **2005**, *53*(6), 2052–2059.

Zhang, H.; Jin, Z. Preparation of Resistant Starch by Hydrolysis of Maize Starch with Pullulanase. *Carbohydr. Polym.* **2011**, *83*, 865–867.

Improvement of Maize: Maize Varieties and Hybrid Maize Technology

ABSTRACT

Maize breeding successfully developed improved maize varieties and hybrids, but it is now facing challenges to develop new varieties with high yield potential, better quality, and resistance to biotic and abiotic stresses. The most significant advances in maize breeding are discussed in this chapter.

11.1 HISTORY OF MAIZE BREEDING

Though maize was accepted as an outstanding forage crop, it was bred only for grain traits for a long time. In the second mid of the 19th century, the first recommendations of maize varieties for specific forage use were given in the French VILMORIN-ANDRIEUX catalogues. In the early 1950s, US hybrids were introduced in Europe which bring about the significant extension of silage maize cropping initiated after the release of early flint x dent hybrids such as INRA258 (1958) and later Brillant DK202, Capella, LG11, and Blizzard G188 (between 1965 and 1975). The extension continued until 1990. In Europe, the first generation of maize hybrids was mostly based on crosses between flint Lacaune and dent Minnesota 13 lines. However, the registration of Dea (1980) in France and Golda in Germany showed remarkable variations in maize dent and to some extent in flint and marked the beginning of a second era of maize hybrid breeding in Europe. The actual maize breeding was initiated after the registration of Banguy (1992), which was characterized by a greater introgression of medium late germplasm into early dent and flint maize lines. During this period, the average genetic improvement in maize yield was nearly 0.10 t/ha/year and reached 0.17 t/ha/year between 1986 and 2004. In early maize, significant improvements of stalk standability, stalk rot, and lodging resistance have been accomplished. The physiological variations related to these improvements were delayed senescence of leaves and stems, higher grain filling rate, and higher stress tolerance. Contrariwise to agronomic value, a constant decline in the cell wall digestibility of hybrids was noted; therefore, in the next future maize should be bred to have a better balance between agronomic and feeding value traits (Barrière et al., 2006).

The adoption of maize hybrids in the US Corn Belt in the early 1930s, increased the maize yields remarkably. This yield gain was achieved with the combined use of plant breeding and improved management practices. On average, about 50% of the increase was due to improved management and 50% due to breeding. The two tools interact so strongly that neither of them could have made such advancement lonely. But, in future years genetic gains may have to bear a larger share of the load. Over the years many variations were observed in hybrid traits. Most frequent changes were noticed in the traits that increases resistance to various biotic and abiotic stresses. However, morphological and physiological variations that stimulate the growth, development, and partitioning efficiency were also documented. Some traits such as grain maturity date in US Corn Belt have not altered over the years because breeders have intended to keep them constant. While, other traits like harvest index have not changed, in spite of breeders' intention to change them. Though breeders have constantly selected for high yield, the necessity to select concurrently for overall reliability has been a driving force in the selection of hybrids with greater stress tolerance over the years. Modern hybrids yield more than their ancestors under both unfavorable and favorable growing environments. The main driving force of the higher yielding ability of modern hybrid was improvement in the capability of the maize plant to overwhelm both biotic and abiotic stresses, rather than increase in primary productivity (Duvick, 2005).

11.2   SUCCESS OF MAIZE BREEDING

During the maize hybrid era viz from 1939 to the present day, commercial grain yields have increased nearly six fold and the genetic component of the improvement has been projected as about 60%. Here, the physiological factors and effective breeding approaches that bring about the yield improvement were examined. Grain yield is the product of accumulating dry matter and allocating a portion of the total dry matter to the grain.

The processes that influence the accumulation of dry matter are normally denoted as the "source" components, whereas the processes that influence dry matter allocation to the grain are denoted as the "sink" components. The variations in leaf canopy size and architecture on the source side account for only a small portion of the improvement. The visual and functional "stay-green" are the major factors of improvement in source capacity. The variations in the association between kernel number per plant and plant growth rate during silking are the major improvement factors on the sink side. In a context of breeding, these advances have been achieved (1) in a "closed" germplasm pool divided into heterotic groups; (2) using the pedigree method of breeding planned to imitate reciprocal recurrent selection and thus improving both additive and nonadditive genetic effects; and (3) by a steady increase in plant population densities during the hybrid era. Functional stay-green and the sink formation dynamics still denote chances for the improvement of yield. However, it is crucial that source and sink should be maintained in balance, and that improvement in one go together with an improvement in the other. Incorporation of high plant population density trials into inbred line development programs is one of the strategies to make use of these opportunities (Lee and Tollenaar, 2007).

To discover and transfer valuable quantitative trait loci (QTL) alleles from unadapted donor lines such as land races and wild species into elite inbred lines, a new method known as advanced backcross QTL analysis was designed. This method combines QTL analysis with the development of variety. In this method, QTL analysis is postponed till the BC_2 or BC_3 generation and all through the development of these populations, negative selection is applied to lessen the occurrence of deleterious donor alleles. Simulations suggested that advanced backcross QTL analysis would be effectual in identifying additive, dominant, partially dominant, or over dominant QTLs. This analysis could detect the epistatic QTLs or QTLs with gene actions varying from recessive to additive with less power than in selfing generations. In one or two additional generations, QTL-NILs representing inbred lines improved for one or more quantitative traits can be obtained from advanced backcross populations. These QTL-NILs can be used to verify QTL activity. The time gap from the discovery of QTL to construction and analysis of improved QTL-NILs is 1–2 years. If effectively applied, advanced backcross QTL analysis can assist in the exploitation of unadapted and exotic germplasm for the improvement of quantitative traits in crop plants (Tanksley and Nelson, 1996).

Maize-teosinte F_2 population was investigated for the genes regulating the morphological differences among maize and its supposed progenitor (teosinte) using molecular markers. Results unveiled the multigenic control of major traits differentiating maize and teosinte. However, few traits like the number of cupules were under the control of a single major locus with a number of modifiers. The other characters like the presence/absence of the pedicellate spikelet exhibited multigenic inheritance. Further, it was found that the tunicate locus (Tu) has no major role in the origin of maize, in spite of the earlier view that it was involved. The major loci contributing to the morphological variances between maize and teosinte were detected on the first four chromosomes. Finally, the study suggested that the variances between teosinte and maize include partly, developmental changes that facilitate the primary lateral inflorescences, which are programmed to develop into tassels (male) in teosinte, to develop ears (female) in maize, and the expression of male secondary sex characters on a female background in maize (Doebley et al., 1990).

A retrospective analysis of the physiological basis of genetic yield improvement may give information about yield potential and may show paths for future yield improvement. The yield improvement of short-season hybrids from the late 1950s to the late 1980s has shown that genetic yield improvement was only 2.5% per year and it was mainly due to the improved stress tolerance. Variances in stress tolerance between older and modern hybrids were observed in terms of high plant population density, weed interference, low night temperatures during the grain-filling period, low soil moisture, low soil N, and a number of herbicides. More efficient resource captures and use results in improved yield. It was found that increased capture of seasonal incident radiation and more uptakes of nutrients and water improves the efficiency of resource capture. Further, the improved resource capture was found to be related with improved leaf longevity, active root system, and a higher ratio of assimilate supply by the leaf canopy (source) and assimilate demand by the grain (sink) during the grain-filling period. Furthermore, maize genetic improvement was found to be accompanied by a reduction in plant-to-plant variability and the results indicated that improved stress tolerance is related with lesser plant-to-plant variability and that higher plant-to-plant variability brings about lower stress tolerance (Tollenaar and Wu, 1999).

Genotyping by sequencing (GBS) technology is an attractive substitute technology for genomic selection. It has the capacity to deliver large numbers of marker genotypes with potentially less ascertainment bias than standard single nucleotide polymorphism (SNP) arrays. But, the utilization of GBS data presents many challenges and the precision of genomic prediction with GBS is presently under study in a number of crops such as maize, wheat, and cassava. In this study, different approaches for incorporating GBS information were evaluated and compared with pedigree models for the prediction of genetic values of lines from two maize populations. Different traits were assessed under diverse environments. Non-imputed, imputed, and GBS-inferred haplotypes of different lengths (short or long) were used to evaluate the methods. Then using Genomic Best Linear Unbiased Predictors (GBLUP) or Reproducing kernel Hilbert spaces (RKHS) regressions GBS and pedigree data were combined with statistical models and accuracy of prediction was measured by means of cross-validation methods. Results revealed that the prediction accuracy of combined pedigree and GBS data was in consistent with the pedigree or marker-only models prediction accuracy. Imputed or non-imputed GBS data showed increased predictive ability than inferred haplotype. The level of prediction accuracy attained with GBS data was similar to those stated by earlier instigators who evaluated this data set using SNP arrays. Under diverse environments, GBLUP and RKHS models combined with pedigree, non-imputed and imputed GBS data provided the best prediction results. However, RKHS gave a somewhat better prediction than GBLUP under drought stressed environments, and both models gave comparable predictions in well-watered environments (Crossa et al., 2013).

11.3   EFFECT OF GREEN EVOLUTION ON MAIZE BREEDING

This section attempts to address the question of whether the "Green Revolution" increased the yield potential of maize hybrids released in the north-central United States. This issue was addressed indirectly by evaluating maize breeding attempts, variations in plant characters of commercial hybrids, and by comparing statewide average yield trends with yield trends in approved yield contests. Based on these sources of information and an explanation of yield potential as the yield that can be realized with an improved hybrid when grown under unstressed conditions, they got conflicting evidences to favor the assumption that maize yield potential has improved. Therefore, the study recommended experimental methods to measure and examine the maize yield potential determinants in the north-central United States and for application in breeding hybrids with high yield potential (Duvick and Cassman, 1998).

11.4   GENETIC DIVERSITY IN MAIZE, GERMPLASM, AND ITS EFFECTIVE UTILIZATION

11.4.1   GERMPLASM POOLS

Two important germplasm pools often utilized in maize breeding are the germplasm pools of Dent and Flint. Quite a number of traits distinguish these two germplasm pools viz cold tolerance, early vigor, and also the flowering time. Unterseer et al. (2016) made a comparative analysis of Dent and Flint genomic architecture that is pertinent for the expression of the quantitative traits. They mentioned that an understanding of the genomic variations that exist between these two germplasm pools may throw in a way to have an enhanced understanding of the complementarity in heterotic patterns for exploitation in the hybrid breeding. It also enhances understanding of the key mechanisms that enable these germplasm pools to adapt to varied environmental conditions.

Almost a century has been passed in the collection and preservation of plant genetic diversity. Germplasm banks that act as storehouses of genetic variation have been set up as a source of genes for improving agricultural crops. The construction of genetic linkage maps made it possible to locate the genes associated with yield and other complex traits on chromosomes. The tools of genome research may lastly release the genetic potential of wild and cultivated germplasm resources for the benefit of society (Tanksley and McCouch, 1997).

Genetic diversity denotes the heritable variation within and among plant species. The genetic diversity within an inter-mating population is the basis for selection and plant improvement. So, conservation of this plant genetic diversity is crucial for current and future human well-being. In recent years, the understanding of the significance of adopting a complete view of biodiversity, which includes agricultural biodiversity conservation for sustainable utilization and development, has been increased. Convention on Biological Diversity and the Global Plan of Action of the Food and Agriculture Organization of the United Nations has protected these principles. Now, the emphasis is to recognize the distribution and amount of genetic diversity accessible to humans in plant species, so as to conserve the genetic diversity and to make its efficient use. It was found that the plant genetic diversity varies with time and space. In a plant species, its evolution and breeding system, ecological and geographical factors, past bottlenecks, and several human factors affects the amount and distribution of genetic diversity. A clear understanding of genetic diversity and its distribution is needed for its conservation and utilization. It assist us in clarifying what to conserve and where to conserve and will increase our knowledge of the taxonomy, origin, and evolution of important plant species. The available resources can be exploited in more valuable ways by improving characterization and by developing core collections based on genetic diversity information (Rao and Hodgkin, 2002).

Association mapping is a powerful tool for studying the molecular basis of phenotypic variations in plants. Association mapping panel of maize with 527 inbred lines from tropical, subtropical, and temperate environments was genotyped by means of 1536 SNPs. The genetic diversity pattern and association among individuals were estimated using 926 SNPs with minor allele frequencies. The analysis unveiled wide phenotypic diversity and complex genetic similarity in the maize panel. Three specific subpopulations were identified in two different Bayesian methods. Later, these subpopulations were reconfirmed by principal component analysis (PCA) and tree-based analyses. The study findings suggested that this maize panel is appropriate for association mapping so as to work out the association between genotypic and phenotypic variations for agriculturally complex quantitative characters with the help of optimal statistical methods (Yang et al., 2011).

Characterization of genetic diversity assists plant breeders in the selection of parental lines and breeding system design. Here, 770 maize inbred lines were screened with 1034 SNP markers and 449 high-quality markers were identified without any germplasm-specific biasing effects. Pairwise comparisons were made among three distinct sets of germplasm, from CIMMYT (394), China (282), and Brazil (94). These comparisons revealed that the elite lines from these diverse breeding pools have been developed with incomplete exploitation of genetic diversity present in the center of origin. Temperate and tropical or subtropical germplasm was distinctly clustered into two distinct groups. The maximum genetic divergence was noted between temperate and tropical/subtropical lines, followed by the variance between yellow and white kernel lines, while the minimum divergence was noted between dent and flint lines. Missing and unique allele's unveiled the substantial levels of genetic variation between diverse breeding pools. Two SNPs obtained from the same candidate gene were found to be related with the divergence between two opposite Chinese heterotic groups. Variation in related allele frequency at two SNPs and their missing allele in Brazilian germplasm showed a linkage disequilibrium block of 142 kb. These findings confirmed the power of SNP markers for diversity analysis and provided a practicable method for the detection and utilization of unique alleles in maize breeding programs (Lu et al., 2009).

Maize is not only a key source of food, feed, and different industrial products but it is also a model genetic organism with vast genetic diversity. Though maize was first domesticated in Mexico, its landraces are extensively found throughout the landmasses. Numerous researches in Mexico and other countries emphasized the genetic variability in the maize germplasm. Applications of molecular markers, especially in the last two decades, provided a better understanding of genetic diversity patterns in maize. This helped in tracing the migration routes of maize from the centers of origin and knowing the fate of genetic diversity during the domestication of maize. In recent years, the genome sequencing of a highly popular US Corn Belt inbred B73 and a popcorn landrace in Mexico, Palomero were significant milestones in maize research. Next-generation sequencing and high-throughput genotyping approaches have the potential to promote our knowledge of genetic diversity and for planning approaches to use the genomic information for the improvement of maize. But, high-throughput and precision phenotyping are the major limiting factor to make use of the genetic diversity in crops like maize. Therefore, there is a crucial need to found a global phenotyping network for complete and effective characterization of maize germplasm for a range of target characters, especially for biotic and abiotic stress resistance and nutritional quality. CIMMYT has initiated a new initiative "Seeds of Discovery" (SeeD) for intensive study of phenotypic and molecular diversity of maize germplasm preserved in the CIMMYT Gene Bank. This initiative is likely to help in efficient documentation and utilization of new alleles and haplotypes for maize improvement. Worldwide multi-institutional efforts are needed to thoroughly investigate the maize germplasm to differentiate the genetic base of elite breeding materials, produce new varieties, and counteract the impacts of global climate change (Prasanna, 2012).

11.5 METHODS OF MAIZE BREEDING

Maize is an important source of food and feed in Asia, and is a source of income for several million farmers. In spite of remarkable advancement achieved in the last few decades via conventional breeding in Asia, average maize yields continue to be low and the demand for maize is predicted to surpass the production in the upcoming years. In the United States and in other places molecular marker-assisted breeding has accelerated the maize yield gains. This method proposes remarkable potential for improving the productivity and value of Asian maize germplasm. The present chapter discussed the significance of such attempts in encountering the increasing demand for maize and provided examples of the latest uses of molecular markers with regard to (1) DNA fingerprinting and genetic diversity analysis of maize germplasm, (2) QTL analysis of major biotic and abiotic stresses, and (3) marker-assisted selection (MAS) for improvement of maize. Further, it highlighted the problems confronted by research institutes desiring to implement the existing and developing molecular technologies. The study concluded that to enhance the level and scope of molecular marker-assisted breeding for maize improvement in Asia, innovative models for resource pooling and intellectual property respecting partnerships will be needed. Scientists should confirm that the molecular marker-assisted breeding tools are determined to develop commercially viable cultivars, improved to ameliorate the most significant constraints in maize production (Prasanna et al., 2010).

Identification of potentially high yielding hybrids, allocating new inbreds to heterotic groups, and determination of the parental line to which a potential donor line is associated are major problems in maize (*Zea mays* L.) breeding programs. In this chapter, the data from a diallel cross among 14 maize inbreds was examined to assess the utilization of molecular markers in solving these problems. For these 14 inbred lines, allozyme genotypes at 14 loci and restriction fragment length polymorphism (RFLP) variant types for 52 cDNA clones were acquired. A measure of genetic distance between inbreds based on marker data viz Modified Roger's distance (MRD) values did not show a significant correlation with hybrid yields. But, a hybrid value based on the number of marker loci with the highest yielding genotype exhibited a significant correlation with hybrid yield. Cluster analysis of MRD values was in agreement with pedigree information but did not show consistency with groupings based on yield. In hybrids with distantly related parents, a relative association measure based on the proportion of homomorphic marker loci showed a significant correlation with a measure of association on the basis of yield (Dudley et al., 1991).

In plant genetics, the application of molecular markers to find QTL influencing valuable characters has become an important method. It helps to understand the genetic basis of these traits and to plan new plant improvement programs. In this chapter, QTLs for seven key traits including grain yield were mapped in two elite maize inbred lines, B73 and Mo17. Two important events in maize genetics-heterosis (hybrid vigor) and genotype-by-environment (G x E) interaction were investigated. Further, two analytical methods of QTL identification viz., the traditional single-marker method and the more recent, interval mapping method were compared. About 3168 maize plots with 100,000 plants cultivated in 3 states were evaluated phenotypically with the help of 76 markers. Both analytical methods provided similar results in the identification of QTLs related with grain yield all over the genome except on chromosome 6. Whenever a QTL related with grain yield was identified, the heterozygote showed a higher phenotype than the relevant homozygote. This suggested that not only overdominance or pseudo-overdominance but also the identified QTLs play a major role in heterosis. This inference was strengthened by a strong association between grain yield and the proportion of heterozygous markers. Though plants were cultivated and evaluated in six different environments, G x E interaction was not evident for most QTLs (Stuber et al., 1992).

An experiment was conducted to assess the grain yield stability of maize hybrids using GGE biplot analysis and Kang's yield-stability statistic (YS_i). Significant variations were noted within years, cultivars, and cultivar-by-location ($C \times L$) interactions. Heterogeneity due to environmental index did not show any significant influence on $C \times L$ interactions. The GGE biplot analysis classified hybrids with above-average yield across years as Hai He > LD10 > YR1 > Tun004 and for stability of performance as LD10, Hai He, Tun004, and YR1. The GGE biplots unveiled that Hai He produced the highest yield in seven and LD10 in 10 locations. GGE biplot and YS_i categorized QC3, XHD892ck, and R313 as the least desirable hybrids. The YS_i showed that ZZY6 and SB21-3 were the most unstable hybrids between years. Overall, YS_i versus GGE distance correlation (r) = −0.92**. Therefore, YS_i could be used to select superior hybrids in the absence of GGE biplot software. This information would be helpful in streamlining the maize testing program in Yunnan (Fan et al., 2007).

In crop plants domestication, selection and breeding are expected to diversify plant lineages and its defenses against herbivores. This study assessed whether defense against corn leafhopper (*Dalbulus maidis*) declines coincidently with life history evolution, domestication, and breeding within the grass genus *Zea* (Poaceae). Four *Zea species* viz., perennial teosinte, Balsas teosinte, landrace maize, and a hybrid maize with three consecutive transitions: the evolutionary transition from perennial to annual life cycle, the agricultural transition from wild annual grass to primitive crop cultivar, and the agronomic transition from primitive to modern crop cultivar were colonized with leafhopper. The development speed, survivorship, fecundity, and body size of leaf hopper were measured. Corn leafhopper exhibited a poor performance on perennial teosinte, intermediate on Balsas teosinte and landrace maize, and greatest on hybrid maize. These results were in agreement with the expectation that defense against leafhopper declines from perennial teosinte to hybrid maize (Dávila-Flores et al., 2013)

11.6 CONCLUSIONS

The maize conventional breeding has been advanced with the application of breeding and new biotechnology techniques to get maize varieties with desired characters like high yield, tolerance to pest and diseases, and better quality. Modern biotechnology enormously improves the precision and decreases the time with which necessary changes in plant features can be made and significantly increases the potential source from which requisite traits can be achieved. In this chapter, the history of maize breeding and its success was discussed.

KEYWORDS

- maize
- improved cultivars
- hybrids
- significant advances
- future objectives

REFERENCES

Barrière, Y.; Alber, D.; Dolstra, O.; Lapierre, C.; Motto, M.; Ordas, A.; Van Waes, J.; Vlasminkel8, L.; Welcker, C.; Monod, J. P. Past and Prospects of Forage Maize Breeding in Europe. II. History, Germplasm Evolution and Correlative Agronomic Changes. *Maydica* **2006**, *51*, 435–449.

Crossa, J.; Beyene, Y.; Kassa, S.; Pérez, P.; Hickey, J. M.; Chen, C.; de los Campos, G.; Burgueño, J.; Windhausen, V. S.; Buckler, E.; Jannink, J. L.; Lopez Cruz, M. A.; Babu, R. Genomic Prediction in Maize Breeding Populations with Genotyping-by-Sequencing. *G3: Genes Genomes Genet.* **2013**, *10*, 1903–1926.

Dávila-Flores, A. M., De Witt, T. J., Bernal, J. S. Facilitated by Nature and Agriculture: Performance of a Specialist Herbivore Improves with Host-Plant Life History Evolution, Domestication, and Breeding. *Oecologia* **2013**, *173*, 1425–1437.

Doebley, J.; Stec, A.; Wendel, J.; Edwards, M. Genetic and Morphological Analysis of a Maize-Teosinte F_2 Population: Implications for the Origin of Maize. *PNAS* **1990**, *87*, 9888–9892.

Dudley, J. W.; Saghai Maroof, M. A.; Rufener, G. K. Molecular Markers and Grouping of Parents in Maize Breeding Programs. *Crop Sci.* **1991**, *31*, 718–723.

Duvick, D. N. The Contribution of Breeding to Yield Advances in Maize (*Zea mays* L.). *Adv. Agron.* **2005**, *86*, 83–145.

Duvick, D. N.; Cassman, K. G. Post-Green Revolution Trends in Yield Potential of Temperate Maize in the North-Central United States. *Crop Sci.* **1999**, *39*, 1622–1630.

Fan, X. M.; Kang, M. S.; Chen, H.; Zhang, Y.; Tan, J.; Xu, C. Yield Stability of Maize Hybrids Evaluated in Multi-Environment Trials in Yunnan, China. *Agron. J.* **2007**, *99*, 220–228.

Lee, E. A.; Tollenaar, M. Physiological Basis of Successful Breeding Strategies for Maize Grain Yield. *Crop Sci.* **2007**, *47*, 202–215.

Lu, Y.; Yan, J.; Guimarães, C. T.; Taba, S.; Hao, Z.; Gao, S.; Chen, S.; Li, J.; Zhang, S.; Vivek, B. S.; Magorokosho, C.; Mugo, S.; Makumbi, D.; Parentoni, S. N.; Shah, T.; Rong, T.; Crouch, J. H.; Xu, Y. Molecular Characterization of Global Maize Breeding Germplasm Based on Genome-Wide Single Nucleotide Polymorphisms. *Theor. Appl. Genet.* **2009**, *120*, 93–115.

Prasanna, B. M.; Pixley, K.; Warburton, M. L.; Xie, C. X. Molecular Marker-Assisted Breeding Options for Maize Improvement in Asia. *Mol. Breed.* **2010**, *26*, 339–356.

Prasanna, B. M. Diversity in Global Maize Germplasm: Characterization and Utilization. *J. Biosci.* **2012**, *37*, 843–855.

Rao, R. V.; Hodgkin, T. Genetic Diversity and Conservation and Utilization of Plant Genetic Resources. *Plant Cell Tissue Organ Cult.* **2002**, *68*, 1–19.

Stuber, C. W.; Lincoln, S. E.; Wolff, D. W.; Helentjaris, T.; Lander, E. S. Identification of Genetic Factors Contributing to Heterosis in a Hybrid from Two Elite Maize Inbred Lines Using Molecular Markers 1992. *Genetics* **1992**, *132*, 823–839.

Tanksley, S. D.; McCouch, S. R. Seed Banks and Molecular Maps: Unlocking Genetic Potential from the Wild. *Science* **1997**, *277*, 1063–1066.

Tanksley, S. D.; Nelson, J. C. Advanced Backcross QTL Analysis: A Method for the Simultaneous Discovery and Transfer of Valuable QTLs from Unadapted Germplasm into Elite Breeding Lines. *Theor. Appl. Genet.* **1996**, *92*, 191–203.

Tollenaar, M.; Wu, J. Yield Improvement in Temperate Maize is Attributable to Greater Stress Tolerance. *Crop Sci.* **1999**, *39*, 1597–1604.

Unterseer, S.; Pophaly, S. D.; Peis, R.; Westermeier, P.; Mayer, M.; Seidel, M. A.; Haberer, G.; Mayer, K. F. X.; Ordas, B.; Pausch, H.; Tellier, A.; Bauer, E.; Schön, C. C. A Comprehensive Study of the Genomic Differentiation Between Temperate Dent and Flint Maize. *Genome Biol.* **2016**, *17*, 137.

Yang, X.; Gao, S.; Xu, S.; Zhang, Z.; Prasanna, B. M.; Li, L.; Li, J.; Yan, J. Characterization of a Global Germplasm Collection and its Potential Utilization for Analysis of Complex Quantitative Traits in Maize. *Mol. Breed.* **2011**, *28*, 511–526.

CHAPTER 12

Research Advances in Breeding and Biotechnology of Maize

ABSTRACT

As maize is an economically and socially significant crop, several advances have been made in its research. Increased applications of new biotechnological methods could bring about an increased rate of improved maize production. Here, the research advances made in maize breeding and biotechnology are reviewed.

12.1 HIGH-YIELDING HYBRIDS

12.1.1 POTENTIALITY OF MAIZE IMPROVEMENT

In a potential commercial hybrid of maize, parents and a collection of elite maize hybrids (6 lines) were resequenced by Lai et al. (2010) revealing more than 30,000 indel polymorphisms and 101 low sequence diversity chromosomal intervals in this maize genome. Apart from the revealing of 100,000 SNPs, many complete genes were revealed. Among these lines, wide variations were detected in the presence or absence of variations in these genes. The complementation of these genes or their mutations (deleterious) in their contribution to heterosis were analyzed and they reported that these high-density SNPs and indel polymorphism markers act as valuable resources in the future molecular genetic studies aimed at bringing about improvements in maize crop.

An experiment was carried out to ascertain the response of maize hybrids from different eras grown widely in Brazil to the increased plant density and to find agronomic features that provide tolerance to high inter-plant competition. The double-cross hybrids, Agroceres 12, Agroceres

303, and the single-cross hybrid Cargill 929, were tested in the main plots. Four plant population densities viz., 25,000, 50,000, 75,000, and 100,000 plants ha^{-1} were applied in subplots. The canopy architecture parameters, accumulation of tassel dry matter, flower synchronization, grain yield, and its constituents were assessed. The higher grain yield was acquired from the older hybrids Agroceres 12 and Agroceres 303 at the lowest plant density. With increasing plant population more drastically increased barrenness, extended anthesis-silking interval and reduced kernel set per ear was observed in the older than in the modern hybrid. The improved performance of modern hybrids at supra optimum plant densities was found to be favored by three sets of traits. Primarily, by lesser partition of dry matter to the tassel, that stimulates a more balanced allometric association between male and female inflorescence. Second, by more compact canopy architecture of modern hybrids, with dwarf plants, fewer and more upright leaves, that enhances interception of solar radiation. Finally, by low ear insertion to plant height ratio that provides better resistance to stalk lodging (Sangoi et al., 2002).

A study was undertaken to assess the genetic gain in total and machine-harvestable grain yield of maize hybrids (Fig. 12.1) and to measure the influence of the plant density: grain-yield interaction on the genetic gain. Nine maize hybrids that had been cultivated during three decades in Ontario were assessed at two locations and four plant densities. At the optimum plant density, the genetic gain in total grain yield was 1.7% yr^{-1}, while for the machine-harvestable grain yield it was 2.6% yr^{-1}. For machine-harvestable grain yield, the optimum plant density was found to increase with the year of hybrid introduction, whereas for total grain yield it increased from old to more recent hybrids. It was found that the reduced stem lodging contributes to about one-third of the genetic gain in machine-harvestable grain yield. The old and modern hybrids comparison indicated that the increased harvest index contributes 15% of the genetic gain in total grain yield (Tollenaar, 1989).

12.2   SUPER SWEET CORN

An experiment was carried out to assess the effects of changes in plant density and nitrogen nutrition on yield and quality of super sweet corn cv. Challenger. Nine plant populations varying from c. 30,000 to 140,000 plants/ha and nitrogen (N) doses, nil and 250 kg N/ha were applied as treatments. Increased grain yield was recorded in plots applied with nitrogen in all the population range. In plots applied with limited N, increment in ear and grain yield was observed with population up to c. 90,000 plants/ha, and then remained constant. More marked effects of limited –N were recorded at high (>90,000 plants/ha) populations. Quality traits like ear size, tip fill, and individual grain mass improved consistently with the addition of N throughout the range of 30,000–140,000 plants/ha. However, the reduction in ear size, tip fill, and individual grain mass was observed with population, irrespective of N supply. Therefore, the study concluded that the optimum balance should be maintained between population and N supply to improve the yield and quality of super sweet corn (Stone et al., 1998).

FIGURE 12.1 Maize hybrid seed production field.

A natural genetic mutant of vegetable corns, viz. waxy or glutinous corn (*Zea mays* L. var. *ceratina*) was found in China. As native waxy corns are open-pollinated with varying ear size, ear shape, kernel color and eating quality, novel hybrids with unique eating quality and appearance can be developed. At Plant Breeding Research Center for Sustainable Agriculture, a waxy corn hybrid breeding program was initiated to develop new varieties with improved eating quality and ear appearance. Two separate base populations were developed by collecting waxy corn cultivars from Thailand and China and super sweet corn cultivars from Thailand and the United States. Inbred lines were obtained and crossed to develop single-cross hybrids. Thus, two single-crossed glutinous corn hybrids viz., white and bicolored hybrids were developed. These were the first glutinous corn hybrids with 75% waxy kernels and 25% super sweet kernels. These hybrids have extra sweetness and delightful eating quality. Both hybrids produced well filled ear with 12–16 rows and can be harvested at about 60 days after planting (Lertrat and Thongnarin, 2008).

A study has been undertaken to find out the optimum harvest time for sweet corn hybrids. Three sweet corn hybrids viz., Great Bell, Danok k1, and Golden Cross Bantam and a super sweet corn hybrid, Crisp Super Sweet 720 were observed for the variations in the sugars and soluble solids contents and flavor rate of cooked kernels from 15 days after silking (DAS) to 27 or 33 DAS. In all hybrids, sucrose content and Crisp Super Sweet 720 fructose and glucose contents increased from 15 DAS to 21 or 24 DAS and then declined. But the highest fructose and glucose contents were recorded at 15 DAS in all the three sweet corn hybrids and then a continuous reduction was observed until maturity. In three sweet corn hybrids, soluble solids increased continuously with maturity, but in crisp super sweet 720 it increased up to 24 DAS, remained at the same level through 30 DAS and then declined. Sucrose was the major sugar at harvest time. At the time of harvest Crisp Super Sweet 720 had 2–3 times higher sugar content and much lower soluble solids than that of three sweet corn hybrids (Lee et al., 1987).

$F_{2:3}$ families of best hybrid of super sweet corn, Jitian 6, were studied for their genetic characteristics of major quality traits. Results revealed the higher broad sense heritability of all the four quality traits. Significant heterosis was recorded for soluble sugar content, protein content, and oil content in F_1 generation. But in $F_{2:3}$ significant heterosis was observed for soluble sugar content and oil content. Soluble sugar content exhibited a significant and positive association with protein content, while it shows a significantly negative correlation with starch content. A significantly, negative association was observed between starch content and protein content. Oil content showed a significant association with protein content, though it did not show any significant correlation with soluble sugar content (Xin et al., 2006).

12.3 DOUBLE HAPLOID MAIZE

In maize (*Zea mays* L.), breeding doubled haploid (DH) lines developed by *in vivo* induction of maternal haploids are normally used. The main advantages of DH lines in hybrid breeding are (1) highest genetic variance, (2) complete homozygosity, (3) short "time to market", (4) simplified logistics, (5) low expenses, and (6) optimum ability for marker applications. This chapter reviewed the experimental basis of the haploid induction technology, elucidated alternate DH-line-based breeding systems, described the characteristics of novel software for improving such systems, and discussed carefully chosen optimization results. Modern inducer genotypes showed average induction rates of 8–10%. Different morphological and physiological markers permit a rapid and cheap identification of haploid kernels or seedlings. Artificial chromosome doubling techniques have been effectively adapted on a large-scale for commercial uses. Probably, haploid embryogenesis is a result of defective sperm cells. The maternal haploids induction rate is under polygenic control. Genome-wide marker-assisted selection can efficiently be combined with DH-line-based breeding approaches. The loss of genetic variation can be lessened by setting lower limits to the effective population size (N_e) so as to maintain selection response in the long run. New software MBP (Version 1.0) exploits the estimated annual genetic gain subjected to budget and N_e limitations. Software input variables contain assessed variance and covariance components, type of tester, haploid induction parameters, and individual breeding activities costs. To calculate N_e, the software takes into account genetic drift caused by both sampling and selection. The optimization outcomes demonstrated that the breeding approaches which involve only one stage of testcross assessment provide faster breeding progress than those with two or more stages and genetic interlinking between staggered breeding programs is more effective than a closed-population method (Geiger and Gordillo, 2009).

The maize haploids produced by using inducer line ZMS were studied. It was found that the tassels of haploid plants were completely sterile. Then the ears fertility was analyzed by pollinating them with the pollen from diploid inbred lines and the resulting crosses had haploids ears with kernels. The haploid plants had an average of 27.4 kernels per ear, which gave rise to normal diploid plants. Owing to this property, the genotypes selected at the haploid plant level could be included in the breeding process. Further, few unusual plants were observed among haploids, these plants phenotypically resembled homozygous lines. It was presumed that the plants might have developed from the spontaneous doubling of chromosomes in haploids (Chalyk, 1994).

The rate of haploid induction is important for *in vivo* DH lines production in maize. Though, the source germplasm utilized as a maternal parent and the induction environmental conditions affects haploid induction. Maternal haploid induction is primarily affected by the inducer used as a pollinator. These characteristics have not been studied in tropical maize until now. In this study, the variation for haploid induction rate (HIR) were observed among different source germplasm in tropical maize, for HIR the relative significance of general combining abilities (GCA) and specific combining abilities (SCA) was determined, and the effect of summer and winter seasons and genotype × season interactions on this HIR was examined. Ten inbred lines were crossed in a half diallele design. The resultant 45 F_1 single crosses were pollinated with the haploid inducer hybrid RWS × UH400 during the summer and winter seasons in a lowland tropical environment of Mexico. The average HIR of the single crosses ranged from 2.90 to 9.66% over seasons. A higher average HIR was recorded during winter. GCA effects of parental inbred lines showed significant variation. However, significant variation was not observed for SCA, GCA × season, and SCA × season interactions. This report highlighted that the selection of suitable source germplasm and pollination under favorable environmental conditions results in higher HIR in tropical maize (Kebede et al., 2011).

12.4 ADVANCES IN MAIZE BIOTECHNOLOGY

To map quantitative trait loci (QTLs), a high-resolution method known as association mapping was developed based on linkage disequilibrium (LD). This method has great potential for the dissection of complex genetic characters. The current assemblage and description of maize association mapping panels, development of advanced statistical methods, and effective association of candidate genes led to the recognition of candidate-gene association mapping power. Though the complexity of the maize, genome presents a number of significant challenges to the use of association mapping, the continuing genome sequencing project will eventually permit a detailed genome-wide analysis of nucleotide polymorphism-trait association (Yu and Buckler, 2006).

Maize (*Zea mays* L.) is a major cereal crop and a model for the study of genetics, evolution, and domestication. To improve the understanding of maize genome organization and to make an outline for genome sequencing, Wei et al. (2007) built a sequence-ready fingerprinted contig-based physical map that covers 93.5% of the genome, of which 86.1% is aligned to the genetic map. The fingerprinted contig map comprises 25,908 genic markers that allowed them to align approximately 73% of the anchored maize genome to the rice genome. The distribution pattern of expressed sequence tags correlates to that of recombination. In collinear regions, 1 kb in rice corresponds to an average of 3.2 kb in maize; however, maize has a six-fold genome size expansion. This can be elucidated by the fact that most rice regions resemble two regions in maize because of its recent polyploid origin. Inversions make up the majority of chromosome structural changes during succeeding maize diploidization. Further, they found clear proof of ancient genome duplication preceding the divergence of the progenitors of maize and rice. Reconstruction of maize genome paleoethnobotany reveals that the progenitors of modern maize had 10 chromosomes.

The two most popular methods that are currently in use for producing transgenic maize are particle bombardment and *Agrobacterium*-mediated transformation. *Agrobacterium*-mediated transformation is likely to give transformants with fewer copies of the transgene and a more predictable pattern of integration. But, when a standard binary vector transformation system is used, these supposed benefits of *Agrobacterium*-mediated transformation trade-off with transformation effectiveness in maize. R_1 and R_2 generations of transgenic maize produced by means of the above two gene delivery methods were compared for the transgene copy numbers and RNA expression levels with the help of southern, northern, real-time PCR, and real-time RT-PCR techniques. The findings revealed lower transgene copies and higher and more stable gene expression in *Agrobacterium*-derived maize transformants. Furthermore, in 70% of transgenic maize produced from *Agrobacterium*-mediated transformation different lengths of bacterial plasmid backbone DNA sequence were found suggesting that the *Agrobacterium*-mediated transformation was not as accurate as earlier observed (Shou et al., 2004).

Genetic diversity is formed by the interaction of drift and selection, but the particulars of this interaction are not well-known. The impact of genetic drift in a population is generally ascertained by its demographic history, usually summarized by its long-term effective population size

(Ne). Fast changing population demographics complicate this association; however, to better understand how varying demography affects selection, Beissinger et al. (2016) used whole-genome sequencing data to study patterns of linked selection in domesticated and wild maize (teosinte). They produced the first whole-genome estimation of the demography of maize domestication, revealing that maize was decreased to about 5% the population size of teosinte before it underwent rapid expansion post-domestication to population sizes much larger than its ancestor. Assessment of patterns of nucleotide diversity in and near genes shows little evidence of selection on valuable amino acid substitutions, and that the domestication bottleneck led to a decline in the effectiveness of purifying selection in maize. Young alleles, however, present proof of much stronger purifying selection in maize, suggesting the much larger effective size of present day populations. The study findings established that the current demographic change-a hall-mark of many species including both humans and crops.

A diverse global collection of 632 maize inbred lines from temperate, tropical, and subtropical public breeding programs was genotyped with Illumina Golden Gate Assay. This assay was developed with 1536 SNPs from 582 loci. Then with the help of 1229 informative SNPs and 1749 haplotypes within 327 loci the genetic diversity, population structure, and ancestral relatedness were estimated. The tropical and temperate subgroups and complex ancestral associations were found within the global collection. Linkage disequilibrium (LD) was determined in overall and within chromosomes, allelic frequency groups, subgroups associated by geographic origin, and subgroups of diverse sample sizes. The LD decay distance varied among chromosomes and ranged between 1 and 10 kb. The LD distance showed increment with the increasing minor allelic frequency (MAF), and with smaller sample sizes, promoting carefulness while using too few lines in a study. Temperate lines had much higher LD decay distance than tropical and subtropical lines because tropical and subtropical lines are more diverse and have more rare alleles than temperate lines. Based on haplotypes, a core set of inbred lines was separated and 60 inbred lines captured 90% of the haplotype diversity. The identified core sets and the whole collection could be extensively utilized in future breeding programs (Yan et al., 2009).

Genetic diversity produced by transposable elements is a key source of functional variation upon which selection acts during evolution. Transposable elements are linked with adaptation to temperate climates in *Drosophila*, a SINE element is related to the domestication of small dog breeds from the gray wolf and there is an indication that transposable elements were targets of selection during human evolution. Though the list of instances of transposable elements linked with host gene function continues to grow, evidence that transposable elements are causal and not just associated with functional variation is inadequate. Studer et al. (2011) showed that a transposable element (Hopscotch) inserted in a regulatory region of the maize domestication gene, teosinte branched1 (tb1), acts as an enhancer of gene expression and partly elucidates the increased apical dominance in maize compared to its progenitor, teosinte. Molecular dating indicated that the Hopscotch insertion predates maize domestication by at least 10,000 years, signifying that selection acted on standing variation rather than new mutation.

The European corn borer (*Ostrinia nubialis* Hübner) is a major pest of maize (*Zea mays* L.). Here, the QTLs influencing resistance against second-generation European corn borer (2ECB) and plant height were mapped and characterized using RFLP markers. B73 (susceptible) × B52 (resistant) were crossed and the resulting 300 F_3 lines were assessed for their parental F_2 genotype at 87 RFLP loci. Plant height and 2ECB resistance data were recorded from the 300 F_3 lines, the parents and the F_2 generation. Plants were artificially infested with 2ECB larvae and damaged stalk tissue was measured to assess the resistance. High heritability values were found for plant height, while tunnel length had intermediate heritability values. The genotypic association between plant height and tunnel length was 0.29. Then, to estimate the genetic effects of these two traits QTLs were localized by using the method of interval mapping. Genomic regions influencing resistance against 2ECB were detected on chromosome arms *1S, 1L, 2S, 2L, 3L, 7L,* and *10L*. Genomic regions affecting plant height were located on *1S, 3L,* and *9L* chromosome arms. The putative QTLs exhibited different types of gene action for both traits. Concurrent mapping of the seven putative QTLs elucidated 38% of the phenotypic variance for tunnel length. Three putative QTLs elucidated phenotypic variance of 63% for plant height. Interaction of genes granting resistance to 2ECB and plant height was evidenced on the *3L* chromosome arm (Schön et al., 1993).

In maize lines, immature zygotic embryos can induce embryogenic callus cultures from which new plants can be regenerated. Here, the transformation of these regenerable maize tissues by electroporation was described. Initially by enzymatic or mechanical method, immature zygotic embryos or embryogenic type I calli were wounded and hen electroporated with a chimeric gene encoding neomycin phosphotransferase (neo).

Transformed embryogenic calli were carefully chosen from electroporated tissues on media containing kanamycin and fertile maize transgenic plants were regenerated. The neogene gets transferred to the progeny of kanamycin-resistant transformants in a Mendelian fashion. This indicated that all transformants were non-chimeric and transformation and regeneration occur in a single cell. This maize transformation procedure does not necessitate the establishment of genotype-dependent embryogenic type II callus or cell suspension cultures and assists the engineering of new characters into agronomically appropriate maize inbred lines (D'Halluin et al., 1992).

MADS-box genes encode transcription factors that are important regulators of plant inflorescence and flower development. Zhao et al. (2011) analyzed DNA sequence variation in 32 maize MADS-box genes and 32 randomly selected maize loci and examined their contribution in maize domestication and improvement. By means of neutrality tests and a test based on coalescent simulation of a bottleneck model, eight MADS-box genes were identified as assumed targets of the artificial selection linked with domestication. As per neutrality tests, one additional MADS-box gene seems to have been under selection during the recent agricultural improvement of maize. For random loci, two genes were designated as targets of selection during domestication and four additional genes were designated to be candidate-selected loci for maize improvement. These results revealed that MADS-box genes were more common targets of selection during domestication than genes selected at random from the genome.

Gene expression variances between different lineages caused by modification of cis-regulatory elements were assumed to be significant in evolution. Lemmon et al. (2014) assayed genome-wide cis and trans-regulatory variances between maize and its wild progenitor, teosinte, using deep RNA sequencing in F_1 hybrid and parent inbred lines for three tissue types (ear, leaf, and stem). Prevalent regulatory variation was detected with about 70% of ~17,000 genes revealing evidence of regulatory divergence between maize and teosinte. However, many genes (1,079 genes) showed constant cis differences with all sampled maize and teosinte lines. For ~70% of these 1,079 genes, the cis differences were limited to a single tissue. Ear tissue had a greater number of genes with cis-regulatory differences, which experienced a drastic transformation in form during domestication. As expected from the domestication bottleneck, maize has less cis-regulatory variation than teosinte, this shortage of genes controlling maize-teosinte cis-regulatory divergence, suggested selection on cis-regulatory differences during domestication. Reliable with selection on cis-regulatory elements, genes with cis effects showed a strong correlation with genes under positive selection during maize domestication and improvement, while genes with trans-regulatory effects did not. They found a directional bias such that genes with cis differences exhibited higher expression of the maize allele more frequently than the teosinte allele, suggesting domestication preferred up-regulation of gene expression. Lastly, this study provides evidence of the cis and trans-regulatory variations between maize and teosinte in over 17,000 genes for three tissues.

In plants, many key regulatory genes that control plant growth and development have been recognized and characterized. In spite of broad information on the function of these genes, much is not known about how they contribute to the natural variation for complex traits. To ascertain whether major regulatory genes of maize contribute to standing variation in Balsas teosinte, Weber et al. (2007) conducted association mapping in 584 Balsas teosinte individuals. They tested 48 markers from 9 candidate regulatory genes against 13 traits for plant and inflorescence architecture. A mixed linear model that regulates multiple levels of relatedness was used to find significant associations. Ten associations comprising five candidate genes were significant after correction for multiple testing, and two endure the conservative Bonferroni correction. zfl2, the maize homolog of FLORICAULA of Antirrhinum, was linked with plant height. zap1, the maize homolog of APETALA1 of Arabidopsis, was linked with inflorescence branching. Five SNPs in the maize domestication gene, teosinte branched1, were significantly linked with either plant or inflorescence architecture. The study findings suggested that major regulatory genes in maize do play a role in the natural variation for complex traits in teosinte and that some of the minor variants detected may have been targets of selection during domestication.

Crop wild relatives are well identified with the reduction in cost of next-generation sequencing and improvement of genomic resources, to bring about major contributions to the field of ecological genomics through full-genome resequencing and reference-assisted *de novo* assembly of genomes of plants from natural populations. The wild relatives of maize, collectively known as teosinte, are a more diverse and characteristic study system than many other model flowering plants. In this appraisal of the population and ecological genomics of the teosintes, Hufford et al. (2012) highlighted recent improvements in the study of maize domestication, introgressive hybridization, and local adaptation, and deliberated future potentials for employing the genomic resources of maize to this interesting group of species. The maize or teosinte study system is an outstanding example of how crops and their wild relatives can bridge the model/non-model gap.

The innovative tool, CRISPR/Cas9 (clustered regularly interspaced short palindromic repeat/CRISPR-associated protein) can be efficiently used to make genome engineering more precise. Owing to its simplicity of use and high specificity CRISPR/Cas9 is a great tool not only for fundamental studies but also for the improvement of important crop traits like higher grain yield, improved tolerance to abiotic and biotic stresses, and better nutritional quality. In this chapter, a step-by-step guide to the CRISPR/Cas9-mediated targeted mutagenesis in maize Hi II genotype was presented. The important steps are designing gRNa, construction of CRISPR/Cas9 vector, transformation of maize immature embryo via *Agrobacterium*, and molecular analysis of transgenic plants to find mutant lines with desirable traits (Lee et al., 2019).

Activation tagging is a powerful tool that distributes transcriptional enhancers all over the genome to induce transcription of adjacent genes so as to determine the function of genes in plants. Here, a transportable element system was developed to allocate a new activation tagging element all over the maize genome. The transposon system was made from the enhancer/suppressor (En/Spm) transposon system that utilizes an engineered seed color marker to detect when the transposon edits. Seed color can detect both somatic and germinal editing actions. The activation tagging element is a *Spm*-derived non-autonomous transposon and has four copies of the sugarcane Bacilliform Virus-enhancer (SCBV-enhancer) and the AAD 1 selectable marker. It was found that the transposon can cause germinal excision events which reintegrates into non-linked genomic locations. The transposon was continued to be active for three generations and actions showing high rates of germinal excision were identified in the T_2 generation. This system generates many activation tagged maize lines that can be screened for agriculturally important phenotypes (Davies et al., 2019).

12.5 CONCLUSIONS

The conventional breeding methods have been advanced with the application of modern breeding techniques and biotechnology tools to acquire plants with desired traits such as improved yield, disease resistance, and better quality. Modern biotechnology tools like genetic engineering and genome editing enormously increased the precision and reduced the time with which desired variations in plant features can be achieved. In this chapter, the current advancements in breeding and biotechnology methods to improve maize are summarized.

KEYWORDS

- **maize**
- **research advances**
- **breeding**
- **biotechnology**

REFERENCES

Beissinger, T. M.; Wang, Li.; Crosby, Kate; Durvasula, A.; Hufford, M. B.; Ross-Ibarra, J. Recent Demography Drives Changes in Linked Selection Across the Maize Genome. *Nat. Plants* **2016**, *2*, 16084.

Chalyk, S. T. Properties of Maternal Haploid Maize Plants and Potential Application to Maize Breeding. *Euphytica* **1994**, *79*, 13–18.

Davies, J. P.; Reddy, V. S.; Liu, X. L.; Reddy, A. S.; Ainley, W. M.; Folkerts, O.; Marri, P.; Jiang, K.; Wagner, D. R. Development of an Activation Tagging System for Maize. *Plant Dir.* **2019**, *3*, e00118.

D'Halluin, K. Bonne, E.; Bossut, M.; De Beuckeleer, M.; Leemans, J. Transgenic Maize Plants by Tissue Electroporation. *Plant Cell* **1992**, *4*, 1495–1505.

Geiger, H. H.; Gordillo, G. A. Doubled Haploids in Hybrid Maize Breeding. *Maydica.* **2009**, *54*, 485–499.

Hufford, M. B.; Bilinski, P.; Pyhäjärvi, T.; Ross-Ibarra, J. Teosinte as a Model System for Population and Ecological Genomics. *Trends Genet.* **2012**, *28*, 606–615.

Kebede, A. Z.; Dhillon, B. S.; Schipprack, W.; Araus, J. L.; Bänziger, M.; Semagn, K.; Alvarado, G.; Melchinger, A. E. Effect of Source Germplasm and Season on the *in vivo* Haploid Induction Rate in Tropical Maize. *Euphytica* **2011**, *180*, 219–226.

Lai, J.; Li, R.; Xu, X.; *et al.* Genome Wide Patterns of Genetic Variation Among Elite Maize Inbred Lines. *Nat. Genet.* **2010**, *42*, 1027–1030.

Lee, S. S.; Kim, T. J.; Park, J. S. Sugars, Soluble Solids and Flavor as Influenced by Maturity of Sweet Corn; Coll. of Agriculture and Animal Science, Yeongnam Univ., Kyongsan (Korea R.), **1987**.

Lee, K.; Zhu, H.; Yang, B.; Wang, K. An Agrobacterium-Mediated CRISPR/Cas9 Platform for Genome Editing in Maize. *Methods Mol. Biol.* **2019**, *1917*, 121–143.

Lemmon, Z. H.; Bukowski, R.; Sun, Q.; Doebley, J. F. The Role of Cis Regulatory Evolution in Maize Domestication. *PLOS Genet.* **2014**, *10*, e1004745.

Lertrat, K.; Thongnarin, N. Novel Approach to Eating Quality Improvement in Local Waxy Corn Kernels from Super Sweet Corn. *Acta Hortic.* **2008**, *769*, 145–150.

Sangoi, L.; Gracietti, M. A.; Rampazzo, C.; Bianchetti, P. Response of Brazilian Maize Hybrids from Different Eras to Changes in Plant Density. *Field Crops Res.* **2002**, *79*, 39–51.

Schön, C. C.; Michael, L.; Melchinger, A. E.; Guthrie, W. D.; Woodman, W. L. Mapping and Characterization of Quantitative Trait Loci Affecting Resistance Against Second-Generation European Corn Borer in Maize with the Aid of RFLPs. *Heredity* **1993**, *70*, 648–659.

Shou, H.; Frame, B. R.; Whitham, S. A.; Wang, K. Assessment of Transgenic Maize Events Produced by Particle Bombardment or *Agrobacterium*-Mediated Transformation. *Mol. Breed.* **2004**, *13*, 201–208.

Stone, P. J.; Sorensen, L. B.; Reid, J. B. Effect of Plant Population and Nitrogen Fertiliser on Yield and Quality of Super Sweet Corn. *Proc. Agron. Soc. New Zealand* **1998**, *28*, 1–5.

Studer, A.; Zhao, Q.; Ross-Ibarra, J.; Doebley, J. Identification of a Functional Transposon Insertion in the Maize Domestication Gene *tb1*. *Nat. Genet.* **2011**, *43*, 1160–1163.

Tollenaar, M. Genetic Improvement in Grain Yield of Commercial Maize Hybrids Grown in Ontario from 1959 to 1988. *Crop Sci.* **1989**, *29*, 1365–1371.

Weber, A.; Clark, R. M.; Vaughn, L.; Sánchez-Gonzalez, J. D. J.; Yu, J.; Yandell, B. S.; Bradbury, P.; Doebley, J. Major Regulatory Genes in Maize Contribute to Standing Variation in Teosinte (*Zea mays ssp. parviglumis*). *Genetics* **2007**, *177*, 2349–2359.

Wei, F.; Coe, E.; Nelson, W.; Bharti, A. K.; Engler, F.; Butler, E.; Kim, H. R.; Goicoechea, J. L.; Chen, M.; Lee, S.; Fuks, G.; Sanchez-Villeda, H.; Schroeder, S.; Fang, Z.; McMullen, M.; Davis, G.; Bowers, J. E. Paterson, A. H.; Schaeffer, M.; Gardiner, J.; Cone, K.; Messing, J.; Soderlund, C.; Wing, R. A. Physical and Genetic Structure of the Maize Genome Reflects its Complex Evolutionary History. *PLOS Genet.* **2007**, *3*, e123.

Xin, Q.; Shuo, J.; Li-juan, C.; Lei, H.; Yu-lan, W. Genetic Analysis of Quality Traits with Super Sweet Corn. J. Jilin Agric. Univ. 2006, 28, 136–138.

Yan, J.; Shah, T.; Warburton, M. L.; Buckler, E. S.; McMullen, M. D.; Crouch, J. Genetic Characterization and Linkage Disequilibrium Estimation of a Global Maize Collection using SNP Markers. *PLOS ONE* **2009**, *4*, e8451.

Yu, J.; Buckler, E. S. Genetic Association Mapping and Genome Organization of Maize. *Curr. Opin. Biotechnol.* **2006**, *17*, 155–160.

Zhao, Q.; Weber, A. L.; McMullen, M. D.; Guill, K.; Doebley, J. MADS-Box Genes of Maize: Frequent Targets of Selection During Domestication. *Genet. Res.* **2011**, *93*, 65–75.

Index

Printed and bound by CPI Group (UK) Ltd, Croydon, CR0 4YY

23/10/2024

01777702-0005